A World Without Time

A WORLD WITHOUT TIME

THE FORGOTTEN LEGACY OF
Gödel AND Einstein

|Palle Yourgrau|

BASIC
BOOKS

A Member of the Perseus Books Group
New York

530.092
Your

Designed by Brent Wilcox
Set in 11 point Sabon

ISBN 0-465-09293-4

CONTENTS

Acknowledgments

Having already written a book intended primarily for philosophers about Kurt Gödel's attempt to make sense of Einstein's theory of relativity, I was intrigued when William Frucht of Basic Books suggested I write another, this one accessible to normal readers. Such a book would focus on the sheer intellectual drama of the companionship of Gödel and Einstein—a relationship sorely neglected in the literature—and would place Gödel's and Einstein's epoch-making discoveries in the context of the great intellectual movements of the twentieth century, some of which, having helped to father, they tried, belatedly, to abandon. It was an offer too good to refuse, and I didn't. The task, however, turned out to be far from easy, and Frucht had to endure not only the late delivery of the final manuscript, but the drumbeat of my complaints about his editorial adventures at the expense of my beloved prose; it may not have been much, but it was all mine. I am grateful to Frucht both for the initial invitation and for (what turned out to be) his wise editorial advice, at every stage, on how to improve the manuscript.

I have greatly benefited from discussing the book with Mary Sullivan and Ben Callard. Sullivan, attentive as ever to the task of trying to keep me honest, read large parts of the book and offered acute, sobering advice, which I took to heart. Callard's critical remarks on every chapter were of great value and are reflected in the final draft. His strange affection for the work, moreover, kept my spirits up during the many dark moments when the project seemed to me ill-advised and misbegotten. Mark van Atten also read the entire manuscript. His extraordinarily detailed remarks, on both substance and style, were a great boon. Robert Tragesser shared with me his deep understanding of Gödel's theorem, and Eli Hirsch helped make certain that my discussions of logic and mathematics were clear, accessible, and to the point. To each of these I owe a serious debt of gratitude, but especially to Callard and van Atten, for their extensive and thoughtful comments.

I would also like to express my appreciation for the conversations I had over a period of years with the late Hao Wang, of Rockefeller University, whom I came to know when I taught at Barnard College in New York. With Wang, who was one of Gödel's closest associates in his final years, I spent endless hours discussing Gödel's ideas, published and unpublished. The question of time was of particular interest to him. He confessed that, finding the topic uncongenial, he had resisted Gödel's efforts to discuss his thoughts on this theme, and regretted it. I regretted it, too. It did not diminish, however, the fruitfulness of our conversations, nor the enjoyment we took in contemplating what Gödel has said and written.

In addition to help from my friends, I have also depended on the kindness of strangers. My copy editor, David Kramer, in addition to improving the clarity and flow of the text, made a number of comments and suggestions of a more substantial nature on a wide variety of topics, from music to mathematics, which were an unexpected gift. AnnaLee Pauls, photoduplication coordinator at the Rare Books and Special Collections department of Princeton University's Firestone Library, provided much needed assistance in the delicate task of choosing and reproducing the right photos from the Gödel archives, housed in Firestone Library. I am grateful both for her assistance and for the spirit with which she provided it. I would also, finally, like to thank Arudra Burra, who is not a stranger, for his generous help, combined with fine judgment, in the search for the right photos to include in the book.

A WORLD WITHOUT TIME

1 | A Conspiracy of Silence

Gödel was ... the only one of our colleagues who walked and
talked on equal terms with Einstein.

FREEMAN DYSON

In the summer of 1942, while German U-boats roamed in wolf packs off the coast of Maine, residents in the small coastal town of Blue Hill were alarmed by the sight of a solitary figure, hands clasped behind his back, hunched over like a comma with his eyes fixed on the ground, making his way along the shore in a seemingly endless midnight stroll. Those who encountered the man were struck by his deep scowl and thick German accent. Speculation mounted that he was a German spy giving secret signals to enemy warships. The dark stranger, however, was no German spy. He was Kurt Gödel, the greatest logician of all time, a beacon in the intellectual landscape of the last thousand years, and the prey he sought was not American ships bound for Britain but rather the so-called continuum hypothesis, a conjecture made by the mathematician Georg Cantor about the number of points on a line. Gödel was spending the summer vacationing at the Blue Hill Inn with his wife, Adele, although fellow visitors at the inn rarely saw either of them. They materialized for dinner, but were never observed actually eating. To the locals, Gödel's scowl betrayed a dark disposition, but the innkeeper saw things differently. For her it was the expression of a man lost in thought. His last word to Blue Hill would not decide the issue. He sent a letter accusing the innkeeper of stealing the key to his trunk.

The place Gödel would return to in the fall was a long way from Blue Hill—the prestigious Institute for Advanced Study in Princeton, New Jersey. There he would no longer have to walk alone, arousing the suspicions of neighbors. He had a walking companion, a colleague at the institute and his best friend. There was no danger that his reputation would intimidate his companion. For his friend, another German-speaking refugee with a mathematical bent, was the most famous scientist of all time, Albert Einstein, whose own meditative strolls already irritated the residents of Princeton.

"From a distance," a biographer wrote, "the [residents of Princeton] chuckled discreetly over [Einstein's] habit of licking an ice cream on Nassau Street on his way home from Fine Hall and were astonished by his utterly un-American long walks through the streets of Princeton." Indeed, toward the end of his career, when he was more or less retired, Einstein commented that his own work no longer meant much to him and that he now went to his office "just to have the privilege of walking home with Kurt Gödel." Ironically, it was not the scowling Gödel but his smiling companion who had once given indirect aid to the German U-boats, when, during World War I, although a courageous and committed pacifist, Einstein had helped improve the gyroscopes used by the German navy. Gödel's research would also, in the end, relate to gyroscopes, but these spun at the center of the universe, not in the dank bowels of submarines.

Washed up onto America's shores by the storm of Nazism that raged in Europe in the 1930s, the two men had awakened to find themselves stranded in the same hushed academic retreat, Princeton's Institute for Advanced Study, an exclusive intellectual club, whose members had only one assigned duty: to think. But Gödel and Einstein already belonged to an even more exclusive club. Together with another German-speaking theorist, Werner Heisenberg, they were the authors of the three most fundamental scientific results of the century. Each man's discovery, moreover, established a profound and disturbing *limitation*. Einstein's theory of relativity set a limit—the speed of light—to the flow of any information-bearing signal. And by defining time in terms of its measurement with clocks, he set a limit to time itself. It was no longer absolute but henceforth limited or relative to a frame of measurement. Heisenberg's uncertainty principle in quantum mechanics set a limit on our simultaneous knowledge of the position and momentum of the fundamental particles of matter. This was not just a restriction on what we can know: for Heisenberg it signified a limit to reality. Finally, Gödel's incompleteness theorem—"the most significant mathematical truth of the century," as it would soon be described in a ceremony at Harvard University—set a permanent limit on our knowledge of the basic truths of mathematics: The complete set of mathematical truths will never be captured by any finite or recursive list of axioms that is fully formal. Thus, no mechanical device, no computer, will ever be able to exhaust the truths of mathematics. It follows immediately, as Gödel was quick to point out, that if we are able somehow to grasp the complete truth in this domain, then we, or our minds, are not machines or computers. (Enthusiasts of artificial intelligence were not amused.)

Einstein, Gödel, Heisenberg: three men whose fundamental scientific results opened up new horizons, paradoxically, by setting limits to thought or reality. Together they embodied the zeitgeist, the spirit of the age. Mysteriously, each had reached an ontological conclusion about *reality* through the employment of an epistemic principle concerning *knowledge*. The dance or dialectic of knowledge and reality—of limit and limitlessness—would become a dominant theme of the twentieth century. Yet Gödel's and Einstein's relations to their century were more uneasy than Heisenberg's.

The zeitgeist took root most famously in quantum mechanics. Here Gödel and Einstein would find themselves in lonely opposition to Heisenberg, who, on the wrong side in the war of nations, chose the winning team in the wars of physics. Heisenberg was a champion of the school of positivism, in quantum physics known as the Copenhagen interpretation in deference to Heisenberg's mentor, the Danish physicist Niels Bohr. What had been a mere heuristic principle in Einstein's special relativity—deducing the nature of reality from limitations on what can be known—became for Heisenberg a kind of religion, a religion that Gödel and Einstein had no wish to join. Some, however, claimed to see in Gödel's theorem itself an echo of Heisenberg's uncertainty principle. That group did not include Gödel.

Einstein, himself one of the great pioneers of quantum mechanics, had known and inspired Heisenberg in Germany. In 1911, in Prague, years before Heisenberg came on the scene, Einstein once pointed out to his colleague Philipp Frank the insane asylum in the park below his study and remarked, "Here you see that portion of lunatics who do not concern themselves with quantum theory." By Einstein's lights, a bad situation had become even worse after Heisenberg. In an early encounter, Heisenberg, on the defensive against Einstein's harangue against quantum mechanics, fought back: "When I objected that in [my approach] I had merely been applying the type of philosophy that he, too, had made the basis of his special theory of relativity, [Einstein] answered simply, 'Perhaps I did use such philosophy earlier, and also wrote it, but it is nonsense all the same.'"

The two parted before the war, Einstein emigrating to the United States, Heisenberg remaining in Germany, to which he would remain loyal to the end. In Princeton, Einstein—pacifist, bohemian, socialist and Jew—was a man apart. To be sure, he found Gödel, but together they remained isolated and alone, not least because of their opposition to Heisenberg's positivist worldview, which ruled the intellectual scene even as Heisenberg's fatherland was attempting to dominate the world. Gödel and Einstein were not merely intellectual engineers, as so many of their brethren, inspired by positivism, had become, but philosopher-scientists. Ironically, while their stars had begun to wane, the sheer size of their reputations had made them unapproachable. Not to each other, however. "Gödel," wrote their colleague Freeman Dyson, "was the only one of our colleagues who walked and talked on equal terms with Einstein."

Their tastes, however, remained distinct. Einstein, a violinist, could never bring his friend to subject himself to the likes of Beethoven and Mozart. Gödel, in turn, had no more success, surely, in dragging Einstein to *Snow White and the Seven Dwarfs*, his favorite movie. History, sadly, does not record which of the seven dwarfs was Gödel's favorite, but we do know why he favored fairy tales: "Only fables," he said, "present the world as it should be and as if it had meaning." That meaning, of course, may be dark. It is not known whether Alan Turing acquired an affection for *Snow White* from Gödel when he visited the institute in the 1930s, but some have heard an echo of the dark side of that tale in Turing's decision to end his life by eating a poisoned apple when, as a reward for his having broken the Enigma code of the German navy, the British government ordered him to receive hormone injections as a "cure" for his homosexuality.

Einstein, before fleeing Germany, had already become a refugee from mathematics. He later said that he could not find, in that garden of many paths, the one to what is fundamental. He turned to the more earthly domain of physics, where the way to the essential was, he thought, clearer. His disdain for mathematics earned him the nickname "lazy dog" from his teacher Hermann Minkowski (who would soon recast the lazy dog's special relativity into its characteristic four-dimensional form). "You know, once you start calculating," Einstein would quip, "you shit yourself up before you know it." Gödel's journey, in contrast, was in the opposite direction. Having befriended Gödel, Einstein commented that he knew now, at last, that in mathematics too one could find a path to the fundamental. In befriending Einstein, Gödel was reawakened to his early interest in physics. On their long walks home from the office, Einstein, forever cheerful, would attempt to raise the spirits of the gloomy and pessimistic Gödel by recounting his latest insights on general relativity. Sadly, however, pessimism blossomed into paranoia. The economist Oskar Morgenstern, calling one day on his good friend, was shocked to find the great Gödel hiding in the cellar behind the furnace.

From those long walks that Einstein and Gödel shared, from their endless discussions, something beautiful would soon be born. The scene was pregnant with possibility. Time, which has taunted thinkers from Plato to Saint Augustine to Kant, had finally met its match in Einstein. While the U-boats of his former fatherland were stalking the Allied fleet, this most un-German of Germans was hunting a more elusive prey. He had amazed the world decades earlier when he alone succeeded in capturing time itself in the equations of relativity. "Every boy in the streets of Göttingen," his countryman David Hilbert wrote, "understands more about four-dimensional geometry than Einstein. Yet, in spite of that, Einstein did the work and not the mathematicians." Relativity had rendered time, the most elusive of beings, manageable and docile by transforming it into a fourth dimension of space, or rather, of relativistic space-time. Sharing with Gödel his latest thoughts on the four-dimensional universe of space-time that he himself had conjured into being, Einstein was sowing the seeds of relativity in the mind of a thinker who would later be described as a combination of Einstein and Kafka.

If Einstein had succeeded in transforming time into space, Gödel would perform a trick yet more magical: He would make time disappear. Having already rocked the mathematical world to its foundations with his incompleteness theorem, Gödel now took aim at Einstein and relativity. Wasting no time, he announced in short order his discovery of new and unsuspected cosmological solutions to the field equations of general relativity, solutions in which time would undergo a shocking transformation. The mathematics, the physics and the philosophy of Gödel's results were all new. In the possible worlds governed by these new cosmological solutions, the so-called rotating or Gödel universes, it turned out that the space-time structure is so greatly warped or curved by the distribution of matter that there exist timelike future-directed paths by which a spaceship, if it travels fast enough—and Gödel worked out the precise speed and fuel requirements, omitting only the lunch menu—can penetrate into any region of the past, present or future.

Gödel, the union of Einstein and Kafka, had for the first time in human history proved, from the equations of relativity, that time travel was not a philosopher's fantasy but a scientific possibility. Yet again he had somehow contrived, from within the very heart of mathematics, to drop a bomb into the laps of the philosophers. The fallout, however, from this mathematical bomb was even more perilous than that from the incompleteness theorem. Gödel was quick to point out that if we can revisit the past, then it never really "passed." But a time that fails to pass is no time at all. Einstein saw at once that if Gödel was right, he had not merely domesticated time: he had killed it. Time, "that mysterious and seemingly self-contradictory being," as Gödel put it, "which, on the other hand, seems to form the basis of the world's and our own existence," turned out in the end to be the world's greatest illusion. In a word, if Einstein's relativity was real, time itself was merely ideal. The father of relativity was shocked. Though he praised Gödel for his great contribution to the theory of relativity, he was fully aware that time, that elusive prey, had once again slipped his net.

But now something truly amazing took place: nothing. Although in the immediate aftermath of Gödel's discoveries a few physicists bestirred themselves to refute him and, when this failed, tried to generalize and explore his results, this brief flurry of interest soon died down. Within a few years the deep footprints in intellectual history traced by Gödel and Einstein in their long walks home had disappeared, dispersed by the harsh winds of fashion and philosophical prejudice. A conspiracy of silence descended on the Einstein-Gödel friendship and its scientific consequences.

An association no less remarkable than the friendship between Michelangelo and Leonardo—if such had occurred—has simply vanished from sight. To this day, not only is the man on the street unaware of the intimate relationship between these two giants of the twentieth century, even the most exhaustive intellectual biographies of Einstein either omit all mention of this friendship or at best begrudge a sentence or two. Whereas a whole industry has grown up in search of Lieserl, the "love child" of Einstein's first marriage, the child of the imagination that was born of the friendship of Einstein and Gödel has been abandoned.

Only in the last few years has this child, the Gödel universe, received any glimmer of recognition. This comes from the redoubtable Stephen Hawking. Revisiting the rotating Gödel universe, Hawking was moved to deliver the highest of compliments. So threatening did he find results like Gödel's demonstrating the consistency of time travel with the laws of relativity, that he put forward what amounts to an anti-Gödel postulate. If accepted, Hawking's famous "chronology protection conjecture" would precisely negate Gödel's contribution to relativity. So physically unacceptable did Hawking find conclusions like Gödel's that he felt compelled to propose what looks like an ad hoc modification of the laws of nature that would have the effect of ruling out the Gödel universe as a genuine physical possibility.

Hawking's attempt to neutralize the Gödel universe shows how dangerous it is to break the conspiracy of silence that has shrouded the Gödel-Einstein connection. Not only does this mysterious silence hide from the world one of the most moving and consequential friendships in the history of science, it also keeps the world from realizing the full implications of the Einstein revolution. It is one thing to overturn, as Einstein did, Newton's centuries-old conception of the absoluteness and independence of space and time. It is quite another to demonstrate that time is not just relative but ideal. Unlike Einstein, a classicist who forever sought continuity with the past, Gödel was at heart an ironist, a truly subversive thinker. With his incompleteness theorem he had shaken the foundations of mathematics, prompting the great mathematician David Hilbert to propose a new law of logic just to refute Gödel's results. The Gödel universe, correctly understood, shares with the incompleteness theorem an underlying methodology and purpose. It is a bomb, built from cosmology's most cherished materials, lobbed into the foundations of physics.

In the footsteps of Gödel and Einstein, then, can be heard an echo of the zeitgeist, a clue to the secret of the great and terrible twentieth century, a century that, like the seventeenth, will go down in history as one of genius. The residents of Blue Hill, preoccupied with war and the enemy out at sea, had failed to take the full measure of their man.

2 | A German Bias for Metaphysics

The German man of science was a philosopher.

J.T. MERZ

It is a remarkable fact . . . that at least in one point relativity theory has furnished a very striking confirmation of Kantian doctrines.

KURT GÖDEL

Physically they were opposites. Gödel, thin to the point of emaciation, hid his spectral body even in the heat of summer in overcoat and scarf. Gaunt, harrowed, and haunted, peering through thick glasses like an owl from another dimension, he could not fail to arouse suspicion. Early in life he had come to the conclusion that the less food one ate the better. This dubious insight he carried out with ruthless consistency, unencumbered by the excess baggage of common sense, a faculty he approached life without. His preconception, fueled by hypochondria that grew out of childhood rheumatic fever and by paranoia about the intentions of doctors, developed into a neurosis that would eventually take his life. During several periods of extreme stress he was confined to sanatoria, from one of which, by some accounts, he enlisted the services of his wife to escape. At his death he weighed a mere sixty-five pounds.

Einstein, in contrast, whose sanity was never in question, was as satisfied by a good sausage as by a good theorem. He had a taste for solid German cooking, which he consumed with relish, topped off by his omnipresent pipe. Friends and wives would be swept aside in the current of his turbulent life, but his pipe never left him. Late in life he was the proud owner of a respectable professorial paunch. "I have firmly resolved," he wrote his wife Elsa, "to bite the dust, when my time comes, with the minimum of medical assistance, and until then to sin cheerfully ... smoke like a chimney, work like a beaver, eat without thought or choice, and walk only in agreeable company, in other words, rarely."

With brown hair and blue eyes, Gödel measured barely five feet six. This number came as a surprise to his colleagues. His intellectual presence was so great that his modest height often went unnoticed. His frailty, however, was obvious. "Of course he has no children," the proprietor of the Blue Hill Inn said of Gödel; "he hasn't the strength to make babies." He did, however, have in his youth the strength to pursue women. "There is no doubt," wrote a college friend, Olga Taussky-Todd, "about the fact that Gödel had a liking for members of the opposite sex, and he made no secret about this fact." Gödel, she went on, was not beyond showing off his acquaintance with a pretty face. Taussky-Todd herself, to her dismay, was once enlisted to come to the mathematical aid of one such young woman who in turn was trying to make an impression on Gödel. Was this interest in women confined to Gödel's youth? Not if his wife, Adele, is to believed. Teasing her husband, she quipped that the Institute for Advanced Study—which she liked to call an *Altersversorgungsheim*, or home for elderly pensioners—was packed with pretty female students who lined up outside the office doors of the great professors. Einstein, who with well-knit limbs and hardy disposition measured five feet nine, did actually make babies, in and out of wedlock. Early and late, the constraints of marriage did not hamper him, even as his discoveries in physics were unconstrained by the conventions of classical physics. The event itself of entering into the institution of marriage bore the unmistakable stamp of unconventionality: Though Einstein wished to marry his cousin Elsa, he desired even more strongly to marry her twenty-year-old daughter, Ilsa. "Albert himself," wrote the flustered daughter to a friend, "is refusing to take any decision; he is prepared to marry either Mama or me."

Clothing too, like marriage, he considered a bourgeois affectation whose strictures he did his best to circumvent, spurning socks, tie, and belt whenever possible. Hair uncut and unkempt, he could embarrass a female guest when his robe, with nothing underneath, fell open, and then express surprise at her consternation. Bursting with the juices of life, he was an indefatigable optimist whose faith in common sense and human nature survived even the Holocaust.

Gödel, in contrast, was in the fullest sense of the phrase "buttoned up." Dressed severely even in the summer heat, he was the very model of dour reserve: gloomy, pessimistic, averse to all human contact except for the closest of friends and the direst of intellectual necessities. The institute still echoes with stories of Gödel's foolproof method for evading a rendezvous. He would carefully arrange a precise location in space and time for the projected meeting. With these coordinates in place, he confided to friends, he had achieved certainty as to where *not* to be when the appointed time arrived. Yet this method had its limitations. Finding himself trapped at an unavoidable institute tea, he negotiated the territory between guests, noted the mathematician Paul Halmos in his memoirs, with maximum attention to the goal of avoiding any possibility of physical contact.

Against every stereotype of the pure mathematician—and particularly one who, like Gödel, had studied and taught in Vienna—Gödel was all but allergic to the masters of classical music, preferring instead light classics and operettas, and was even more so to the abstractions of modern art. He was untouched by intellectual snobbery and made plain his love of fairy tales. His fondness for Walt Disney cartoons was no secret to his friends. Comedies, however, he disliked.

Einstein was consumed by his passion for the great Austrian-German classicists, Bach, Mozart and Beethoven, but especially Mozart. His friend and biographer, Philipp Frank, offered some shrewd observations about what made Mozart special. What passed for many as a sign of Einstein's cynicism was for Frank an expression rather of Einstein's urge "to make the serious things in the world tolerable by means of a playful guise." But this also characterizes much of Mozart's music, "which might also be called 'cynical.' It does not take our tragic world very seriously." Einstein was always ready to perform Mozart at a moment's notice on his beloved violin, which he played, myths notwithstanding, very well. He "was an experienced sight reader," wrote the professional violinist Boris Schwartz, "with a steady rhythm, excellent intonation, a clear and pure tone, and a minimum of vibrato." Only his pipe was as familiar a companion. Violin and pipe: these will be forever the icons of the great scientist, together with his tousled hair.

Gödel, as is clear from photographs, was meticulously clean-shaven, every hair combed in place, whereas as every schoolchild knows, a small brush of a moustache floated above Einstein's full lips. Combs, moreover, in the Einstein household were verboten. With the visual signature comes the acoustic: When something or someone struck Einstein funny, a huge belly laugh welled up inside the scientist and erupted like a volcano that shook his entire body. More than a few times it shook up as well the surprised object of this laughter, who realized too late its full meaning. Gödel, in contrast, had a soft, high pitched chuckle, more a musing to himself on the ironies of the universe than a full-throated laugh. Raising the pitch of his voice at the end of each sentence and trailing off into silence, he left his audience with a feeling of detached query. (As a child of four he had been nicknamed *Herr Warum*, Mr. Why. "Why is your nose so large?" he asked an embarrassed guest.)

By age a generation apart, Einstein and Gödel shared an anniversary by one degree of separation. The year of Einstein's birth, 1879, was that of Gödel's mother, Marianne. (It was also the year that saw publication of Gottlob Frege's masterpiece, *Begriffsschrift,* and thus the birth of modern mathematical logic, a field Gödel would raise to unparalleled heights.) They were born into different religions, Einstein a Jew, Gödel baptized a Lutheran. Skeptical of the faith of his fathers in his youth, with the rise of Nazism Einstein rediscovered what he called his "tribal companions" and became a passionate, if thorny, Zionist. He never did, however, embrace the transcendent God of his people, accounting himself rather a "deeply religious unbeliever." His hero was not Moses but Spinoza, the pantheist and excommunicant, and he reflected this predilection throughout a scientific career in which such seemingly transcendent, untouchable things as space, time and light were revealed to be fully immanent and subject to physical causality.

Gödel was not a pantheist but rather a self-described theist, "following Leibniz," he said, "not Spinoza." Spinoza's God, he said, "is less than a person. Mine is more than a person. . . . He can play the role of a person." He noted the oft-neglected fact that the founders of modern science were not atheists. More radical than Einstein, he belonged to a rare breed of thinker: the true believers. Whereas "ninety per cent of philosophers these days," he would say, "consider it the business of philosophy to knock religion out of people's heads," he would exploit the machinery of modern logic to reconstruct Leibniz's famous "ontological argument" for the existence of God. Though not a Jew, he was nevertheless taken for one. In a Vienna teeming with Nazis, his wife once employed her umbrella to fend off a group of rowdies who were jostling Gödel, mistaking him for a Jew.

The misattribution was not confined to Nazis. While at the Institute for Advanced Study in Princeton, Gödel was for a time a member of an elite—a very elite—discussion group, consisting of himself, Einstein, the German physicist Wolfgang Pauli, and Bertrand Russell, one of the founders of modern "analytical" philosophy. Russell reacted badly to the discussions, finding them too philosophical in the "old-fashioned sense." (The failings of an entire century are crystallized in this fact.) In an unpleasant aside he vented his frustration: "All three of the others were Jews and exiles, and in intention, cosmopolitans," he wrote later, "[who shared] a German bias for metaphysics." "I am not a Jew," Gödel would respond later, "even though I don't think this question is of any importance." He admired the tenacity of the Jewish people. "Kurt had a friendly attitude toward people of the Jewish faith," said his friend Olga Taussky-Todd. "And once he said out of the blue that it was a miracle how, without a country, they were able to survive for thousands of years, almost like a nation, merely by their faith." Einstein, wishing to eliminate the Jewish need for miracles, pushed hard for most of his life for a homeland for the nation that had survived so many years without a home. Never too concerned with consistency—unlike his logician companion—he was undisturbed by his earlier briefs against nationalism.

Seeing the handwriting on the wall, Einstein and Gödel abandoned comfortable university positions in Berlin and Vienna when the Nazis came to power in the 1930s. At the zenith of their powers, they were snatched up by the newly formed Institute for Advanced Study in Princeton, good European root stock for the vineyards of the new world. Together they wandered the narrow streets of a cloistered and provincial academic town, they who once strode the boulevards of the great capitals of Europe, centerpieces of a once great civilization crashing down in ruins. Strangers to each other in Europe, it was not until 1942 that they began the friendship that lasted until Einstein's death in 1955, a loss from which Gödel never recovered. Einstein, a German Jew in a nest of Wasps, felt out of place in Princeton. Gödel, already a recluse, resented less the isolation, although his wife, Adele, suffered. A café dancer in Vienna, Adele was out of her element in the elite college town. When an opportunity opened up to move to Harvard, she pleaded for the more cosmopolitan Cambridge, Massachusetts. But Gödel was not prepared to accept an offer where teaching was required.

What attraction could have drawn together such opposites as Einstein and Gödel? Certainly not scientific agreement. This was not a case of the strong force uniting like-charged protons in the atomic nucleus. The charges here were opposite. Gödel opined, in fact, that one of the reasons Einstein enjoyed his company was precisely because he made no attempt to hide his very different views, not just in politics and philosophy but in physics. "I frequently held an opinion," Gödel said, "counter to Einstein's and made no attempt to conceal my disagreement." Einstein's failed search, for example, for a unified field theory to unite the domain of quantum mechanics with general relativity, which occupied much of their discussions, was a particular target of Gödel's skepticism.

Indeed, Gödel was skeptical of the ultimate significance of natural science itself, despite its great success in enabling us (as he put it) to build TVs and bombs. At a faculty dinner at the institute the young John Bahcall, having introduced himself as a new astrophysicist on the faculty, was taken aback when Gödel replied flatly that he didn't believe in natural science. By Gödel's lights, physics had taken the wrong turn centuries ago when it chose to follow the path laid by the naturalistically minded British empiricist Isaac Newton, rather than that of the German idealist Gottfried Leibniz. Gödel's fascination with Leibniz was boundless, prompting a mathematical colleague, Paul Erdös, to offer a rebuke: "You became a mathematician," he told Gödel, "so that people should study you, not that you should study Leibniz." Gödel even succeeded in transferring his own paranoia to Leibniz, arguing at length that some of his hero's crucial manuscripts had been secretly destroyed by "those who do not want man to become more intelligent." "You have a vicarious persecution complex," replied his friend Karl Menger, "on Leibniz's behalf." Menger, like most intellectuals a child of the Enlightenment, went on to ask why none of Voltaire's papers had been destroyed. "Who ever became more intelligent," Gödel answered, "by reading Voltaire?"

Further separating Einstein from Gödel was the fact that Einstein never fully resolved his native suspicion of mathematics. To the end, the great physicist favored his cherished physical intuitions. Even though it was precisely Minkowski's mathematical reworking of special relativity in terms of four-dimensional geometry (which Einstein resented at the time) that led to the mathematical abstractions of general relativity, the physicist remained forever wary of being led by the nose by mathematicians. He confessed once to being suspicious of a new move in general relativity that he said he could reach only mathematically (i.e., not intuitively). Gödel, in contrast, always felt most secure when he had formulated a problem in symbolic, mathematical terms. "If you had a particular problem in mind," wrote Taussky-Todd, "he would start by writing it down in symbols." Yet Gödel also believed, famously, that in mathematics too there are intuitions (a doctrine for which logicians still have not forgiven him). For Gödel the equations of mathematics, as opposed to the counsels of common sense, would lead us into the promised land of new insights, whereas for Einstein, it was precisely common sense that was the final touchstone for assessing what the mathematicians had to offer.

Beneath these disagreements, however, or beyond them, there was much that united the two minds. Both had grown to maturity in the ancient capitals of Europe. They were heirs to the great Austrian-Germanic philosophical tradition, with "philosophy" understood here in its widest sense. Prejudice aside, Russell's comment on the "German bias for metaphysics" had not really missed its mark. Raised in this culture, the composer Gustav Mahler had kept, quite naturally, in his "composing hut," volumes by both Wolfgang Goethe and Immanuel Kant. It comes as no surprise, then, that Gödel and Einstein cut their philosophical teeth on the great works of Kant, whose fingerprints can be clearly discerned throughout the work of each. For Gödel, his writings on Einstein were as much an expression of his interest in Kant's and Leibniz's ideas of time as of his personal association with Einstein. He would characterize his own contributions to relativity theory—to Einstein's consternation—as showing that relativity had "verified" Kant's philosophical idealism.

Einstein's own reading of Kant, in turn, did much to free him from the excessive reliance on immediate sensory data to which many of his contemporaries, especially Ernst Mach, were susceptible. At the tender age of sixteen Einstein had reread Kant's weighty masterpiece, *The Critique of Pure Reason*—the same age at which Gödel too read Kant—and as a student at the Technical Institute in Zurich he had enrolled in a course on Kant. Still, he often made light of the tendency, especially strong in Germany, to venerate the German master. "Kant," he said, "is a sort of highway with lots and lots of milestones. Then all the little dogs come and each deposits his bit at the milestones."

"At the Institute in Princeton," Gerald Holton has noted, "[Einstein's] favorite topic of discussion with his friend Kurt Gödel was . . . Kant." Kant, deeply impressed by Newton—much of his *Critique*, indeed, was intended to provide a philosophical foundation for Newton and Euclid—had made famous the doctrine that science is fundamental and rigorous exactly to the degree to which it is mathematical. Einstein and Gödel, in turn, each in his own way, approached the world mathematically. For both, mathematics was a window onto ultimate reality, not, as for many of their scientific colleagues, a mere tool for intellectual bookkeeping.

Huddled over a desk in Fine Hall or walking home from the institute, they were a model of mathematical companionship. A chance photograph taken by a visiting mathematician finds the two friends together on the road, each sporting a white straw hat, Einstein beaming for the camera, his convex body bursting from rumpled, baggy pants held up by an ancient pair of suspenders, while the white linen of Gödel's fitted coat holds him closely, his eyes fixed in a cold stare. (Two gentlemen farmers from a Faulkner novel, commented one observer.) Each had found in the other a rare companion who could resist the charms of the "new physics" of Bohr and Heisenberg, according to whom mathematics could no longer provide for science a picture of the world as it actually is in itself—a worldview—but could serve only as a tool for calculation, a means for predicting the outcome of experiments. An impossible prescription to follow for "Mr. Why," and no less so for Einstein. For a signature of Einsteinian science is the Socratic search for "definitions," for what something "really is," in itself (a favorite expression of Plato's). Einstein, after all, was the man who had taught Kant what time "really was" (the fourth dimension of relativistic space-time), taught Newton what gravity "really was" (the curvature of four-dimensional space-time), and taught everyone what energy "really was" (as every schoolchild knows, $E = mc^2$).

As students of Kant, Einstein and Gödel were well aware that although space and time are the two fundamental forms of human experience—space, as Kant had it, the form of intuition of "outer sense," time the form of "inner sense"—it was space that was the natural object of scientific inquiry. And it was space that was first captured by the Greek mathematician Euclid, whose axiomatic-deductive system of geometry—the bane of every high school student—became the paradigm of science, a model from Newton to Einstein. Even in his new physics of space, Einstein had simply generalized geometry from Euclid to the new non-Euclidean geometries, in which the angles of a triangle could sum to less, or more, than 180 degrees. (To the end of his life, Einstein could wax nostalgic about a boyhood gift that had turned his life around, his "holy geometry booklet.")

Yet as Einstein and Gödel well knew, it is not space but time that in the end poses the greatest challenge to science. The dynamic nature of time, the fact that it flows, is obviously its most striking feature. But it is another thing entirely to make sense of this seemingly obvious truth. After all, to flow is to flow in time. What sense can one attach, then, to the idea of the flow of time itself? Saint Augustine, in his *Confessions*, tied himself in knots over such conundrums. Western thought as such, one might say, is characterized by a kind of geometrical Midas touch. Whatever science touches becomes subject to geometry, the science of space. "Time," Kant himself had said, "is nothing but the form of inner sense, that is, of the intuition of ourselves and our inner states . . . and just because this inner intuition yields no [geometrical] shape, we endeavor to make up for this want by analogies." The analogy, for Kant, is to think of time, which is not space, as spatial! "We cannot," said Kant, "obtain for ourselves a representation of time which is not an object of outer intuition [i.e., of sensory experience] except under the [spatial] image of a line."

Thus when Einstein in 1905 captured time in special relativity, he once again transformed it into space, this time, into the fourth, temporal component of the geometrical structure of four-dimensional "space-time." Not for nothing did G.J. Whitrow write, "the primary object of Einstein's profound researches on the forces of nature has been well epitomized in the slogan, 'the geometrization of physics,' time being completely absorbed into the geometry of a hyper-space." The universe, however, not being empty of matter, is not governed by the matter-free idealization of special relativity but rather by Einstein's next brainchild, the general theory of relativity, the subject of Einstein's free tutorials with Gödel. Worse, the world of general relativity, much to Einstein's displeasure, was actually "expanding," that is, expanding over time. (God, apparently, had for once failed to consult first with Einstein.)

But special relativity had taught the world that simultaneity, and thus time, is not, as Newton thought, worldwide and absolute, but rather local and relative. In what sense of time, then, could the universe itself be expanding, absolutely, over time? Time itself must have been smiling over the puzzle it had created. Appearances notwithstanding, Einstein had not after all succeeded in trapping this elusive prey in the net of general relativity. As Hubble showed, the universe really is expanding! The problem could not be avoided. But if even Einstein had run aground on these rocky shoals, who was left to take the lead? Whom could one compare with Einstein if not his traveling companion in general relativity, Kurt Gödel? But what made Gödel the logician, whose universe consisted of the timeless mathematical realm of sets and numbers, the right person to carry forward Einstein's torch into the uncertainties of the new space-time?

3 | Vienna: Logical Circles

After one session in which Schlick, Hahn, Neurath and Waismann had talked about language, but in which neither Gödel nor I had spoken a word, I said on the way home, "Today we out-Wittgensteined these Wittgensteinians: we kept silent."

KARL MENGER

Born into the Austrian-German minority of Brno, a city now in the Czech Republic, the place where Mendel laid the foundations of the science of genetics, the Gödel brothers, Rudolf and Kurt, took it as a given that they would undertake their final academic studies at the storied University of Vienna. Vienna remained even after the Great War one of the premier intellectual centers of the world, distinguished in law, medicine (Rudolf would become a radiologist), physics, mathematics, social sciences, economics, philosophy, and theology. In those years there passed through the city many of the individuals who created the twentieth century, including Sigmund Freud, the founder of psychoanalysis; the composers Richard Strauss and Gustav Mahler as well as Arnold Schoenberg, the inventor of twelve-tone music; the painters Gustav Klimt and Oscar Kokoschka, as well as the revolutionary architect Adolf Loos, who presaged the famous Bauhaus school; the physicist-philosophers Ludwig Boltzmann and Ernst Mach; and the philosophers Karl Popper and Ludwig Wittgenstein. The list could be extended indefinitely. Wittgenstein, himself a kind of minimalist, was an admirer of the minimalism practiced by Loos, and harbored architectural designs of his own. The attraction was mutual: "You are me!" said Loos to Wittgenstein when they met in 1914. The ratio of intellectual genius to square footage in Gödel's Vienna takes one's breath away.

Among those who were privileged to think the unthinkable, however, there is another name that belongs here. Adolf Hitler's path to Vienna began in Linz, the city of his birth, where in 1904 he attended the same realschule as Wittgenstein. Though the same age as young Ludwig, young Adolf was two years behind him at school. There exists a class photograph in which Wittgenstein appears to be placed near Hitler.

Both of the Gödel boys excelled in secondary school in Brno, but Kurt's gifts were clearly exceptional. He was a standout in all subjects, from science and mathematics to languages, and is said never to have made a single error in his Latin exercises. (It was in mathematics, ironically, that he received his only less than perfect grade.) Arriving in Vienna in 1924, Gödel decided at first to concentrate in physics, a choice that would serve him well. He also received a solid grounding in philosophy, especially the history of philosophy, with Heinrich Gomperz, and excelled in all his classes in mathematics, a subject in which he acquired by graduation a remarkable degree of depth as well as breadth, from geometry to number theory and mathematical logic. It would soon emerge that he was embarked on an intellectual journey in the direction of increased rigor and precision, from mathematical physics to mathematics, from there to mathematical logic, and finally from mathematical logic to mathematical philosophy.

As an undergraduate, Gödel was particularly impressed by the lectures on number theory, attended by hundreds of students, given by Philip Furtwängler, a cousin of the legendary orchestral conductor Wilhelm Furtwängler, whose fame in those years would turn to infamy when he declined to leave Germany during the next world war. Gödel claimed later that Philip Furtwängler, who was paralyzed from the neck down, gave the best lectures he had ever heard. It was Furtwängler whom Gödel credited with his turn to mathematics. The drama of his lectures was heightened by the fact that Furtwängler lectured from his wheelchair, without notes, while an assistant wrote equations on the blackboard. The feeling of disembodiment this engendered fit the subject of the lectures perfectly. The natural numbers 0, 1, 2, 3, . . . seem to possess the kind of independent existence and "geometry" usually reserved for concrete physical objects. Unsurprisingly, therefore, this branch of mathematics is a breeding ground for Platonists, who like Plato believe in the objective, independent existence of ideal, disembodied "forms," of which the natural numbers are a paradigm. These are no more subject to the arbitrary manipulations of the human will than the distant stars, which we observe but cannot touch. As the minimalist mathematician Kronecker put it, "God made the natural numbers; all else is the work of man." For Gödel, all numbers are "the work of God."

Gödel's journey from physics to mathematical logic took place just as the new field was coming into its own as a well-established intellectual enterprise, although, truth be told, logic remains to this day in the eyes of many mathematicians a poor relation, not quite mathematics, not quite philosophy. Having for centuries been the province of rhetoricians and grammarians, logic emerged as a branch of mathematics at the turn of the century, due in large part to the work of the German philosopher-mathematician Gottlob Frege, an acquaintance of both Russell and Wittgenstein and a seminal influence on their thinking. Frege's early masterpiece, *Begriffsschrift* (*Concept Script*), published in 1879, succeeded in simultaneously axiomatizing logic and formalizing it, that is, formulating it in an artificially constructed, purely symbolic language, prefiguring today's computer programming languages. The rules of such a language are unambiguous and can be followed "mechanically," without the need to understand the meaning of the symbols. Not content with this, Frege employed this new mathematized logic—which for him was not a mere calculating device, but a proper science, with its own content and subject matter—as itself a foundation for mathematics, in particular for arithmetic, or number theory.

Another German mathematician, David Hilbert, in turn developed a mathematical theory devoted to the study of the new symbolic logical systems like Frege's, a field that came to be known as metamathematics, since it involved the mathematical study of mathematical systems themselves. (Hilbert had been introduced to the idea of metamathematics—and more generally to the notion of a metalanguage—by the Dutch mathematician L. E. J. Brouwer when they shared a holiday in 1909 at the resort of Scheveningen, the Netherlands. In the course of time, Brouwer would become Hilbert's nemesis.) This went beyond Frege, who believed that there is only one genuine logical system, the one he had developed, and that there was no "stepping outside" it to compare it with other systems or with the objects themselves, i.e., mathematical models—a conception outside the mathematical mainstream that would eventually be taken up by Frege's admirer Wittgenstein. It was indeed in this new mathematical field of metamathematics that the question of completeness was raised by Hilbert, who asked whether a given symbolic logical system, such as the one developed by Frege in *Begriffsschrift,* given its axioms and proof procedures, was both internally consistent (the axioms and proof procedures could not be used to prove two statements that contradicted each other) and complete (the proof procedures sufficed to prove every true statement in the system). It was in answering just such questions that Gödel discovered his famous incompleteness theorem.

The development of mathematical logic also went beyond Frege insofar as it replaced Frege's use of concepts (the *Begriffe* of his *Begriffsschrift*) with the extensions of these concepts—i.e., the set of objects described by the concepts—which came to be known as sets or classes. But whereas for Frege the theory of concepts and their extensions had been contained in logic itself—the very part of Frege's theory that Bertrand Russell would later show contained an inconsistency—as the field developed in the early years of the twentieth century, set theory came into being as a new field unto itself, with its own axioms. This new axiomatic set theory, developed by such thinkers as Ernst Zermelo and Abraham Fraenkel, replaced both the axiomatized theory of concepts and their extensions of Frege, as well as the earlier, unaxiomatized, "naïve" set theory of Cantor. Frege's work gave birth, then, to two new subfields of mathematics: mathematical logic and axiomatic set theory.

Gödel's mathematical advisor, Hans Hahn, kept abreast of all these new developments in mathematical logic and set theory. Indeed, he directed a seminar devoted to the classic of modern mathematical logic, *Principia Mathematica*, by Bertrand Russell and Albert North Whitehead. Gödel did not participate in that seminar, but he did attend one given by the philosopher Moritz Schlick on Bertrand Russell's later work, *Introduction to Mathematical Philosophy* (written while Russell was in jail in England for his protests against British participation in World War I). Gödel also attended a seminar on the foundations of mathematics offered by Rudolph Carnap, who had been a student of Frege's at the University of Jena. Carnap would shortly become one of the most influential members of Schlick's "Vienna Circle." In a city full of cliques, salons and discussion groups on every conceivable topic, the Vienna Circle was the most exclusive.

Outside the circle, Gödel's life was not unlike those of other well-off Viennese intellectuals. With his brother, Rudolf, his senior by four years, Kurt lived in a comfortable apartment in which they were joined regularly by their mother, Marianne. Together with her, the two brothers enjoyed automotive vacations in Rudolf's new Chrysler, one of the first in the region, to spots as far away as Marienbad. Though the family employed a chauffeur, on vacations the brothers preferred doing the driving. Kurt liked to drive fast. This, combined with his penchant for indulging in abstract reverie while behind the wheel, led his future wife, Adele, to put an end to his driving career. In town, Marianne made certain that her two academic sons did not neglect the full cultural offerings laid before them by beautiful Vienna. These included plays, with box seats at Max Reinhardt's famous Josefstadt Theater, and concerts, especially light opera, of which Gödel was especially fond.

Gödel's exceptionally clear mind made him a much sought after intellectual companion among his fellow students. He was generous with his time and patient with his interlocutors. His friend Karl Menger writes that Gödel "always grasped problematic points quickly and his replies often opened new perspectives for the enquirer. He expressed all his insights . . . with a certain shyness and a charm that awoke warm and personal feelings for him in many a listener." But Gödel's shyness should not be mistaken for timidity. When the already distinguished Carnap suggested to his young student that he write some encyclopedia entries to gain recognition, Gödel responded that he had no need for such devices to achieve renown. Nor was he timid with women, or above a little showing off. His fellow student Olga Taussky-Todd would later describe one particular encounter that impressed her, albeit negatively. In a classroom near the mathematical seminar, "the door opened and a very small, very young girl entered. She was good-looking . . . and wore a beautiful, quite unusual summer dress. Not much later Kurt entered and . . . the two of them left together. It seemed a clear show off on the part of Kurt." Later, this same young woman sought the reluctant Taussky-Todd's mathematical assistance in an attempt, apparently, to impress Gödel. (She complained to Taussky-Todd, however, about Gödel being spoiled, inclined to rise late in the morning, and so on.) According to Rudolf, his brother developed a particular fondness for a family-run restaurant near their apartment, an attraction he attributed to Kurt's interest in the attractive twenty-year-old daughter, who served as a waitress.

Generally, however, Gödel seems to have preferred the company of older women. His very first romantic interest, the daughter of friends who used to visit his family, was described as an "eccentric beauty," but she was also ten years his senior, and his parents put an end to the relationship. More serious was the attachment he formed at the age of twenty-one for Adele Porkert, a nightclub performer—self described as a "ballet dancer," a profession at that time only marginally more acceptable—employed at *Der Nachtfalter* (The Moth). Six years his senior, Adele was married, her face partially disfigured by a "port wine stain," and Catholic (a religion for which neither Gödel nor his parents had any sympathy). Her marriage, however, was brief and unhappy, and a prolonged romance ensued between the dancer and the mathematician. Georg Kreisel, who used to visit Kurt and Adele, described her as lacking formal education but possessing "a real flair for *le mot juste*." In addition, according to Kreisel, Adele liked to tease Kurt by constructing "far-fetched grounds for jealousy," and also by making fun of his curious interest in ghosts and demons. Kurt's parents, for reasons that are obvious if not admirable, were not amused by Adele and objected strenuously to the relationship. It was only after his father's death in 1929, at the age of fifty-four, that marriage to Adele became a possibility, though it took nine more years for the deal to be clinched. The delay, according to Gödel's friend in later life, Hao Wang, may have been partly responsible for the fact that the Gödels never had children, a circumstance that would weigh heavily on the increasingly sad and lonely Adele.

The brothers were separated not only by Kurt's interest in women—his brother never married—but by the fact that while Rudolf spent the day at the hospital, Kurt attended the university. If not at the university or with a woman or attending a play, he would repair frequently to one of Vienna's famous coffeehouses, generally to one of the cafès that were the preferred haunts of the Vienna Circle, such as the *Reichsrat, Schottentor,* or *Arkaden*. Though far from gregarious, Gödel developed close friendships with several colleagues and professors, including Carnap and Hahn, as well as Herbert Feigl and Marcel Natkin of the Schlick circle. Von Neumann, too, became a friend, and a close one at that. Feigl, for his part, recalled long walks with Gödel through the parks of Vienna and coffeehouse discussions of matters philosophical, logical, mathematical, and scientific that continued late into the night. And Karl Menger, who was reported to have been the favorite student of Gödel's advisor, Hahn, also became a good friend of Gödel's, inviting him to participate in—and eventually edit the proceedings of—the mathematical colloquium he founded. The most organized and regular philosophical interactions, however, between Gödel and other minds were doubtless the weekly discussions conducted in Schlick's Vienna Circle, of which he became a regular member in 1926, having been introduced into the circle by Hahn.

Gödel's Vienna was a city of coffeehouses, each devoted to a particular intellectual theme—those with white table tops, convenient for writing formulas, being especially favored by mathematicians—as well as of intellectual circles, especially philosophical ones. The theme of the Vienna Circle was logical positivism. Though a guest in the house of Schlick, Gödel was hardly enamored of the circle's credo of positivism, nor of the hero of this cult, Ludwig Wittgenstein. The bible of the Vienna Circle was Wittgenstein's *Tractatus* (*Tractatus Logico-Philosophicus* was the full title, suggested by Wittgenstein's friend and former teacher, G. E. Moore, emulating Spinoza's *Tractatus Theologico-Politicus*), completed while the author was a prisoner of war. But Wittgenstein's true war, like that of the Vienna Circle, was not against the Allies but against metaphysics. Positivism, a particularly severe brand of intellectual minimalism—a spirit that thrived in Gödel's Vienna—is an antiphilosophical philosophy dedicated to the belief that most of what has passed for deep metaphysical thinking over the centuries is nothing more than confusion based on an inadequate understanding of language, which, through artifice, leads the mind by the nose in all the wrong directions.

Gödel did not share the positivist credo that philosophy begins and ends with an analysis of language and its limitations, nor Wittgensteinian's doctrine that the subject matter of traditional philosophy, as opposed to that of physical science, is precisely that which cannot be expressed in language. He had no sympathy for the famous line with which the *Tractatus* concludes, that "what we cannot speak about we must pass over in silence," as shown in a reminiscence by Menger after the two had attended a session of the Vienna Circle: "Today we . . . out-Wittgensteined these Wittgensteinians; we kept silent." Apparently, Gödel and Wittgenstein never met, though Gödel said that he saw him once, when both attended a lecture in Vienna by the Dutch anti-Platonist, "intuitionist" mathematician, L.E. J. Brouwer.

In the meetings of the Vienna Circle, Gödel rarely spoke, signaling his agreement or disagreement only by a slight inclination of the head. Participation in these meetings was by invitation only, and membership hovered between ten and twenty. The regular participants included Schlick and Carnap, the philosophers Carl Hempel, Otto Neurath, Friedrich Waismann and Feigl, and finally, Menger, Hahn and Gödel. Conspicuous by their absence were the philosophers Popper and Wittgenstein, the former because he had not been invited due to his views about the latter, the latter because he had declined the invitation. The meetings took place in a dingy room filled with rows of chairs and long tables on the ground floor of the building in the *Boltzmanngasse* that housed the mathematical and physical institutes. Early arrivals were expected to clear the chairs away from the blackboard to allow the day's speaker room to maneuver. One table was reserved for smokers. Intellectually, the circle was devoted to the theme of positivism, the doctrine that physical science, whose ultimate basis is sensory experience, exhausts what can be known, leaving philosophy the task primarily of policing the ever-present tendency of thought to pretend to more knowledge than can be delivered by science. Wittgenstein himself, though their hero, was not a positivist. What separated him from them was this: what must be "passed over in silence" was for Wittgenstein precisely what had value.

The note struck by the positivists was hardly new. Immanuel Kant had declared centuries earlier that "reason," as such, stands in need of an internal "critique." In the "Dialectic" of his *Critique of Pure Reason* he described in detail reason's attempts to fly through thin air—a region, he noted, heavily populated by philosophers. Nevertheless, Kant himself proceeded to develop a system of philosophy that pretended to a kind of knowledge not derivable from science. This the new positivism rejected. What gave it its "logical" twist were the recent efforts by Frege, Russell, Hilbert, and others to develop logic both as an instrument that served to formalize the physical sciences—and thus to assist in their policing—and as a new branch of mathematics that was simultaneously a foundation for the rest of mathematics and a close cousin to what was worth preserving in the philosophical tradition. It was unsurprising that Schlick's logical positivists chose as their patron saint the Wittgenstein of the *Tractatus*, since it was one of the themes of that slender but potent work that a primary task of philosophy is to separate clearly and forever what can be said from what cannot, to clear out centuries of philosophical clutter and render a clear path for science.

For Wittgenstein, the new logic of Frege and Russell provided the tool that made not just the attempt but the completion of this task possible: "I . . . believe myself to have found," he declared modestly in his preface, "on all essential points, the final solution of the problems." Wittgenstein, moreover, had a line on a problem that had haunted the positivists. The physical sciences that served as their model for all thought were rigorous precisely to the degree that they were mathematical, yet mathematics itself is not a physical science. It appears altogether immune to the touchstone of sensory experience that forms the very basis of physical science. Without an account of mathematics, then, the new minimalist edifice of logical positivism threatened to crumble under its own weight. The *Tractatus* was a gift from God, for if Wittgenstein was right, mathematics was not a science at all. Strictly speaking, there is no proper knowledge in mathematics, no truth. Rather, the systems of equations represent conventional rules for the manipulation of abstract symbols that make possible the genuine knowledge offered by physical science. With this approach, mathematics as such is merely a calculus, a calculating device, not a language of thought, as it was for Frege. As Frege's former student Carnap put it, mathematics is not a genuine language that can express thoughts but rather the "logical syntax of language." This was a doctrine that Gödel, the true heir to Frege, would spend the rest of his life working to defeat.

Wittgenstein was the patron saint of the Vienna Circle. "I can testify to this . . . ," wrote Olga Taussky-Todd. "Wittgenstein was the idol of this group. . . . An argument could be settled by citing his *Trac-tatus*." The unofficial saint, however, was Gödel's future friend Einstein, considered by many to be the greatest scientist since Newton. (Wittgenstein himself once stated that in a sense he was a follower of Einstein.) It was not only that Einstein's theories had revolutionized the scientific image of the world. The philosophical aspects of Einstein's theory of relativity held a special appeal. In the special theory of relativity, Einstein had rejected Newton's "metaphysical" postulates of absolute space and time, which resisted any direct empirical confirmation. Time, Einstein insisted, was physically real only to the extent that it could be measured by a clock. (As the positivists would put it, the meaning of a term consists in its method of verification.) Since physical experimentation demonstrated that not all clocks could be definitively synchronized, Einstein declared that time was not after all absolute, as Newton had believed, but rather relative to the frame of reference of the clock by which it was measured. Similarly, since there existed no definitive empirical method to detect whether an object's motion through space was absolute, Einstein declared that all spatial relations were also relative to a given reference frame chosen, by convention, as the "rest frame." For the positivists, the success of this theory meant that the tenets of their credo made for good science, while their rejection could lead to bad philosophy or a scientific dead end.

For Einstein, the rejection of absolute space and time was merely a statement about the physical world. It was much more, however, to the Vienna Circle. As much a religion as a scientific methodology, positivism denied that science or philosophy, with the help of mathematics, could provide a *Weltbild*, or worldview, a picture or account of ultimate reality. They could only supply the scientist with a method of calculation and prediction. In addition to rejecting the task of providing a metaphysical picture of reality—the very goal of philosophy itself as it has been historically practiced—the positivist confined his epistemology to direct sensory experience, rejecting any claims for insight into, or intuition of, the concepts of the theorist or the abstract objects of the mathematician. Gödel and Einstein, in contrast, held the faculty of intuition in the highest regard. "I put my faith in organization," John D. Rockefeller Jr. once said on meeting Einstein. "I put my faith in intuition," the physicist replied. Gödel was even more explicit. In perhaps his most (in)famous philosophical remark, he laid down the gauntlet against positivism: "Despite their remoteness from sense-experience, we do have something like a perception of the objects of set theory, as is seen from the fact that the axioms force themselves upon us as being true. I don't see any reason why we should have less confidence in this kind of perception, i.e., in mathematical intuition, than in sense-perception." For the positivist, however, there is neither an abstract realm of concepts nor a human faculty of intuition that could provide insight into this realm. The concepts invoked by the philosopher must give way to the techniques employed by the engineer. Thus Wittgenstein, a sometime aeronautical engineer, in the *Tractatus*: Mathematics is not, as Frege had it, a science of the platonic realm of mathematical concepts and objects, but rather a system of techniques for the manipulation of mathematical signs.

The War of the Titans

That it was Vienna that gave birth to this extreme antiphilosophical philosophy was no accident, nor was it an accident that the Vienna Circle was its cradle. Moritz Schlick, the founder of the Circle, had in 1922 assumed the chair in the philosophy of the inductive sciences formerly occupied by Ludwig Boltzmann, and before that Ernst Mach. The philosopher-physicist Mach was a prominent figure at the University of Vienna, occupying the chair in the history and philosophy of the inductive sciences from 1895 to 1901. In 1864, he had been professor of mathematics at Graz, in 1867, professor of physics at Prague. He made important contributions to acoustics, electricity, hydrodynamics, mechanics, optics and thermodynamics. It was Mach who in 1887 laid the basis for the principles of supersonics, which demonstrated that a material object traveling past the speed of sound would have an effect that is now called a "sonic boom." An object's speed relative to the speed of sound is today called its "Mach number," Mach 2, for example, indicating a velocity of twice the speed of sound.

Mach had an enormous influence on his contemporaries. He was an early devotee of an extreme version of positivism and was a passionate advocate for his doctrines. His polemics succeeded in making a generation of scientists skeptical not only of theorists' speculations about the microscopic world, but even of the extended use of abstract mathematics as an element of physical theories. He was a successful polemicist and popularizer, and he acquired a kind of cult following among intellectuals of various stripes. His admirers included the young poet Hugo von Hofmannsthal, a member of the artistic circle Young Vienna, who would become famous as the librettist for the operas of the composer Richard Strauss.

Mach advocated a "critique," somewhat in Kant's sense, of physical science, attempting to purge it of all elements not verifiable by sensory experience. "Pseudoproblems" he called the traditional concerns of philosophy (a term for which Carnap developed a strong attachment); "antimetaphysical" he subtitled the introductory remarks to one of his books. For Mach, the ultimate foundation of science was the data offered by the five senses. He was ruthless in his rejection of any conceptions that resisted empirical confirmation. Newton's absolute space and time were for him anathema. Is the sun "really standing still" while the earth revolves around it, or is it the earth that is "really stationary"? The question for Mach was nonsensical, since physical science contains no detector for one's position in "absolute space." Einstein, clearly, was impressed. He cited Mach explicitly as a seminal influence on his special theory of relativity. But Mach's influence on him was primarily negative, clearing the way for Einstein to find, with no help from Mach, a positive theory to replace Newton's. "Mach's way," wrote Einstein to his old friend Michele Besso, "cannot give birth to anything living; it can only exterminate harmful vermin."

Mach's other maxims could prove positively harmful to an attempt to find a successful scientific theory. For one, he was constitutionally opposed to the forming of "models" in physics, simplified abstract simulacra of real, complex phenomena, fruitful in forming theoretical constructions that can be tested against empirical facts. He labored under the illusion that science could be built upon the basis of "inductions," generalizations based on patterns in observed phenomena. Einstein's proclivity was for gedankenexperiments, thought experiments, in which the imagination manipulated images to see what would happen if a hypothesis not necessarily derived from sensory experience were true. He was more sympathetic, however, to another of Mach's maxims, against the excessive use of abstract mathematics in theoretical physics. Mach's preoccupation with the data of sensory experience made him suspicious of high flights of the mathematical imagination. He suspected mathematicians of substituting the artful manipulation of symbols for the honest work of empirical testing and confirmation. Einstein was sympathetic. The mathematics of his revolutionary paper on special relativity was relatively elementary, and at first he resisted its reformulation in terms of four-dimensional space-time by his former teacher Hermann Minkowski, complaining that "since the mathematicians pounced on the relativity theory I no longer understand it myself." Unlike Mach, however, he quickly came to his senses, and fortunately so, since the progression from special to general relativity—Einstein's crowning achievement—would have been impossible without Minkowski's mathematical reformulation.

Einstein believed that physical reality contained more than what we can derive from the data of sensory experience. The real world, for him, was what corresponded to physical theory. It consisted of entities like atoms and force fields, in themselves undetectable by the senses, but indirectly discernible by their effects on systems that can affect human or artificial receptors. Mach, in contrast, had no truck with theoretical constructs and remained skeptical of anything that could not be reduced to a law based on a combination of sensory experiences. This bias had several consequences, each individually devastating to scientific inquiry. First, Mach remained to his final days violently opposed to the new scientific view that much of the real world consists of entities, like atoms, forever invisible to the unaided human senses. This stance by such a powerful figure hindered scientists, in their research and in their careers, who did not share Mach's prejudice, and ensured that Mach's own scientific worldview would become increasingly irrelevant. As Einstein wrote later, "the antipathy of these scholars [Ostwald, Mach] towards atomic theory can indubitably be traced back to their positivistic philosophical attitude. This is an interesting example of the fact that even scholars of audacious spirit and fine instinct can be obstructed in the interpretation of facts by philosophical prejudices."

Second, Mach, by intention the most empiricist of thinkers, was rendered indistinguishable from the philosophical "idealists" who believe that the real world is simply a fiction created by the human mind. The sensations that for Mach formed the very basis of science are after all individual, private, subjective mental phenomena that cannot be shared, in direct contrast to the shared, objective, independent physical world that constitutes the mainstay of the empiricist worldview. Ironically, it is the mathematical Platonist Frege who turns out to be a greater empirical realist than the supposedly hardnosed Mach. In "Thought," a late essay, Frege wrote, "We really experience only [our mental] ideas, not their causes. And if the scientist wants to avoid all mere hypothesis, then he is left just with ideas; everything dissolves into ideas, even the light rays [and] nerve fibers . . . from which he started. So he finally undermines the foundations of his own construction." Frege was not the only prominent figure to be alarmed by Mach's tendency toward idealism. In *Materialism and Empirio-Criticism,* in 1909, V. I. Lenin took time out from the revolution to launch a spirited critique of Mach's idealistic tendencies. That the busy Lenin thought it necessary to refute him is an indication of Mach's reach.

In direct opposition to Mach stood his contemporary Ludwig Boltzmann, the founder of the statistical theory of mechanics. A gifted pianist, Boltzmann loved to play Beethoven's symphonies in their piano transcriptions by Franz Liszt. He was also fond of Wagner and at one point took piano lessons from the composer and organist Anton Bruckner. The lessons came to an abrupt end when Boltzmann's mother discovered a wet raincoat the composer had left on the bed. Boltzmann had an unstable personality but a warm and soft heart. Short and stocky, with thick eyeglasses, curly hair, an equally curly beard, and a surprisingly high-pitched voice, he was called by his fiancée "my sweet fat darling." Mach felt otherwise. The enmity between the two was such that after Mach accepted the chair in the philosophy of the inductive sciences in Vienna in 1895 and began giving lectures to large enthusiastic audiences, Boltzmann resigned the chair of theoretical physics and moved to Leipzig, where he immediately encountered another scientific enemy in Wilhelm Ostwald. Depressed by continual arguments with Ostwald, Boltzmann attempted suicide. He was pleased to return to Vienna in 1901 when Mach, having received an appointment to the Austrian parliament, resigned the chair in philosophy, leaving it open for Boltzmann. When Boltzmann began lecturing on philosophy, his classes became so popular that the university's largest lecture room could not contain them.

It was Boltzmann who introduced probabilistic thinking as essential to physics. The behavior of the molecules of a gas could be seen to obey precise laws only to the extent to which their motions were considered in the aggregate, statistically. It is one of the great paradoxes of physics that the principles of mechanics are time-symmetric—they operate identically if run in reverse—and yet disorder increases with time. Boltzmann was able to demonstrate that the second law of thermodynamics—that an independent physical system always moves over time in the direction of "maximum entropy" (i.e., maximum disorder)—holds if the system is given a statistical or probabilistic interpretation. Since the direction of maximum entropy has been held to account for the "direction of time" and for the existence of irreversible physical processes like breaking an egg, Boltzmann's contribution was enormous. It was certainly not lost on the young Einstein, who embraced probability eagerly, and it later paved the way for the use of probabilistic methods in quantum mechanics (which an older Einstein found far less palatable).

Again in opposition to Mach, Boltzmann was a firm believer in the importance of mathematical methods in the physical sciences, pointing the way to the future of physics, while Mach faced toward the past. In mathematizing a physical theory, not only did one escape the idealism and subjectivity of Mach's "sensationalism," one could also use the power of mathematical manipulation itself to achieve new insights that would never have been discovered through mere empirical observation. Boltzmann, unlike Mach, believed in the power of theoretical models to increase the scope of physical theory, and yet again in opposition to Mach, he held that these models, if successful, establish a genuine conception of the world that describes physical reality itself— a realm, like the world of microscopic atoms, invisible to the unaided human senses.

A war ensued between the titans Mach and Boltzmann over whether atoms were genuine features of the physical world or merely useful fictions formulated to assist the physicist. Though a powerful and dynamic teacher, Boltzmann was an inferior polemicist compared to Mach, who carried the day on that battlefield, whether in person or in print and whether in Vienna with Boltzmann or waging battle from afar. In 1897, in one particularly unpleasant encounter at a meeting of the Imperial Academy of Sciences in Vienna, Mach, following Boltzmann's talk, rose to voice his objections, declaring bluntly, "I don't believe that atoms exist!" Boltzmann never did recover from such polemics, in spite of the fact that in 1905 Einstein would deliver a terminal blow to Mach's epistemological critique of the atomic hypothesis.

In the same year, indeed, in the same volume of the same journal, *Annalen der Physik*, in which he propounded the theory of relativity, Einstein published a paper on "Brownian motion," the random dance long observed in the behavior of microparticles, in which he made explicit use of the hypothesis of the reality of atoms to explain this phenomenon and to make precise and verifiable predictions about the behavior of particles based on measurements of real, nonfictional, atoms. In the same year, Einstein had managed to make profound physical discoveries employing both Mach's verificationist critique of Newton (to produce the theory of relativity) and Boltzmann's mathematical and model-theoretic realism (to produce the theory of Brownian motion and establish the existence of invisible atoms). This was neither the first nor the last time Einstein would succeed in overthrowing not just the theories but the very worldview of a great thinker who had once been his inspiration or father figure. He would write years later that "the scientist must appear to the systematic epistemologist as a type of unscrupulous opportunist; he appears as a realist insofar as he seeks to describe a world independent of the acts of perception . . . as a positivist insofar as he considers concepts and theories justified only to the extent to which they furnish a logical representation of relations among sensory experiences."

Boltzmann's rescue by Einstein, however, came too late. Worn down by years of verbal battles with the ruthless Mach, Boltzmann succumbed to his wildly unstable moods. In 1906, on holiday at the Bay of Duino near Trieste, the great physicist tried once again to take his own life, while his wife and daughters were swimming, this time by hanging. This second attempt succeeded. A great influence, not only on Einstein, had come to a sudden and tragic end. Ludwig Wittgenstein, who had been hoping to make a career in physics and engineering by following Boltzmann's great model, now embarked on a new course that would ultimately lead him, via research in aeronautical engineering in Manchester, England, to Frege and Russell and the foundations of mathematics. Boltzmann's influence, however, had already left its mark. The physicist's prophetic idea of describing a physical system by locating it in a logical framework in various dimensions of physical significance would have a profound effect not only on the future of quantum mechanics but on the bible of the Schlick circle. For it was in Boltzmann's conception that Wittgenstein found the germ of his idea of locating an object—any object—by its position in what in the *Tractatus* he called "logical space." Where Boltzmann chose to locate a physical body by specifying its position in terms of a set of co-ordinates, three spatial and one temporal, as well as a fifth coordinate, temperature, and a sixth, pressure, and so on, which gave the "ensemble of possible states" of a physical system, Wittgenstein wrote that "the facts in logical space are the world. . . . A picture presents a situation in logical space. . . ."

It was no accident that Wittgenstein's *Tractatus* was required reading in Schlick's Vienna Circle of philosophically minded scientists and scientifically minded philosophers. Moritz Schlick himself had been a student of Max Planck, the seminal figure in quantum mechanics. He was deeply impressed by Einstein's theory of relativity and published extensively on it, as well as on general epistemology. His most influential work was probably *The General Theory of Knowledge*. He was clear, quiet, and soft-spoken, politically liberal like most of his associates, but not as likely as they were to mix politics with science and philosophy. He did not strike Karl Menger as a German from Berlin. Yet, when asked by Menger at a party if he was really from Berlin, Schlick's answer was, "Sad, but true." Schlick tended to idolize intellectual figures, though only, as Menger put it, "figures of the first order." A fascination with Einstein was succeeded by one for Russell, to be succeeded in turn by an even stronger worship of Wittgenstein. It was in the Vienna Circle that Schlick tried to bring together the strands of physics, philosophy and mathematics that had emerged in recent decades.

Unlike Boltzmann, Schlick was hardly an exciting speaker. His lectures, delivered in a barely audible monotone, were characterized more by precision than by passion. Silver-haired and sporting an elegant vest, he was a model of sober dignity. Philosophically, he rejected Mach's skepticism about the existence of sense-transcendent entities like atoms, cleaving, rather, to Boltzmann's line. But he agreed with Mach that verification is the lifeblood of physical theories, and together with Carnap, Hempel and others, he raised the methodological maxims that had guided Einstein's first steps in relativity to the level of a formal theoretical postulate: the meaning of a scientific term is exhausted by its method of verification (the "verifiability theory of meaning"). The Vienna Circle made a serious and honest attempt to come to grips with the extraordinary developments that had taken place at the turn of the century in physics, philosophy and mathematics, but under Schlick's guidance, positivism reigned supreme, even if not in the crude form it had taken under Mach. It was precisely the hegemony of positivism, Gödel wrote later, that allowed the members of the circle to mistake Einstein for an ally and to underestimate the difficulty of rendering mathematics empirically acceptable by reconstructing it as a system for the formal manipulation of signs. Einstein himself would awaken the positivists from their misconceptions about the ultimate relationship between his thought and theirs. And Gödel, in short order, would surprise everyone by striking a fatal blow to the most rigorous attempt to reconstitute mathematics as a formal theory of signs.

Schlick himself was never fully awakened. In 1936, when he was fifty-four years old, a mentally unstable former doctoral student of his named Hans Nelböck began to hound and threaten him. Nelböck had been spurned by a fellow student, Sylvia Borowicka, whose romantic inclinations were reserved for the leader of the Vienna Circle. Whether her affections were reciprocated is not known. In addition, Nelböck's attempts to find work had come to nothing, something he also held against Schlick, whose complaints against him had been the original source of his treatment for mental disorder. Nelböck would show up in class, glaring over his spectacles at the mild professor, and later make menacing phone calls when the lecturer had retired to his home. Deeply concerned, Schlick notified the police and acquired a bodyguard, but to no avail. Early on the morning of June 21, 1936, the deranged student encountered him on the steps of the philosophy building and with an automatic pistol fired four rounds point blank, killing him instantly. "Now you damned bastard, there you have it!" he is reported to have screamed as he stood over Schlick's body, the gun smoking in his hand. Thus was the Vienna Circle abruptly closed, its history framed by a suicide and a murder.

Long considered a haven for atheism, communism, materialism and assorted other crimes of its Jewish professors, Schlick's circle had become a target for the increasingly virulent strain of anti-Semitism in Austrian nationalism. Schlick's death was taken to be a promising development. "It is to be hoped," wrote one newspaper, "that the terrible murder at the University of Vienna will quicken efforts to find a truly satisfactory solution to the Jewish question." The writer, apparently, did not know or care that Schlick was neither Jewish nor an atheist but rather a German Protestant. He had, in fact, but one Jewish assistant, Friedrich Waismann, who had already been dismissed as part of the university's attempt to rid itself of Jews. Nelböck's sentencing upon conviction was a rather lenient ten years, whereas hanging was the customary penalty. The court cited his mental instability as a mitigating factor. He was, however, forced to sleep on a hard bed, with a new one delivered every three months. After the Anschluss, Nelböck became a kind of folk hero. He was released on probation and spent the war as a geological technician for the Third Reich. At long last, Nelböck had found work.

The Meaning of Relativity

What Schlick had been seeking in the Vienna Circle was a unified epistemology—a systematic account of what can really be known—on the basis of a philosophically coherent interpretation of what Einstein had achieved in physics and what Frege, Russell, Hilbert and their predecessors had achieved in the foundations of mathematics. What exactly had Einstein accomplished, however, and what was the meaning of the new direction in the foundations of mathematics? As the century was turning, a perceived tension had arisen between Newton's laws of motion—according to which measurement should always be relative to, and affected by, the state of motion of the observer—and the equations of the Scottish physicist James Clerk Maxwell, one of Einstein's heroes, who had unified the theories of electromagnetism and optics.

Maxwell's equations gave as the speed of light a velocity, denoted by c, relative to a reference system at rest in Newton's absolute space, known otherwise as the ether, a substance whose existence was theorized from first principles rather than empirical evidence. This invariance suggested that one could determine whether a given reference system was at rest or in motion relative to the ether by testing the speed of light relative to this system. Any deviation from c would signal motion relative to absolute space. All efforts at measurement, however, such as those performed in the famous (because exquisitely precise) Michelson-Morley experiments, failed to detect any variation in the speed of light, which seemed impossible: since the reference frames tested were in motion relative to each other, they could not *all* be at rest in the ether! Amazingly, although bullets shot from a moving train have a velocity that is increased by the speed of the engine, the measurable velocity of a light beam sent out from this same engine is unaffected by the train's speed. The classical principle of "addition of velocities" was in peril.

Into the breach stepped the Dutch physicist Hendrik Antoon Lorentz, Einstein's father figure and another of his heroes. It was Lorentz who, having already perfected the form of Maxwell's equations, appeared now to save the day by supplying the exact equations, the "Lorentz transformations," that made measurements in one reference, or inertial, frame equivalent to those obtained in another, including the "absolute" rest frame of the postulated ether. In one stroke, Lorentz had succeeded in crystallizing and rendering harmless the mismatch between Newton's account and Maxwell's. Once the Lorentz transformations were applied, the same physical laws could be seen to hold in the absolute rest frame of the ether and in any other inertial frame.

For most serious thinkers, including Lorentz himself, these "transformations" signaled at most our inability to measure the "real" velocity of light or our real velocity through Newton's postulated ether of "absolute space." It was assumed that there were simply unavoidable distortions in the measurement of light. For the young Einstein, in contrast, the mismatch demonstrated that space and time themselves—what Kant had called the fundamental "forms of intuition"—needed to be re-created or redefined. Henceforth, with the Einstein revolution, time itself, not just its measurement with clocks, would be understood as something essentially relative to the motion of the observer and his or her frame of reference, as something in its essence related to the speed of light. Instead of trying to explain our inability to detect the "true" speed of light, then, Einstein incorporated that speed into the very definition of time and motion. Lorentzian engineering had been replaced by Einsteinian metaphysics.

In a move that contained as much philosophy as it did science, Einstein had succeeded in combining the letter of positivism—rejecting any properties of space and time that could not be determined through measurement with rods and clocks—with the spirit of German metaphysics, determining what kind of things space and time are. Though generations of physicists, not least the redoubtable Heisenberg, would conclude that Einstein had become the self-appointed standard-bearer of positivism, the truth lay elsewhere. He was, if anything, an opportunist.

Einstein was merely exploiting, for his own philosophical purposes, certain elements of positivism that in the particular case of special relativity were justified. Gödel too would come to exploit elements of the positivist methodology—in his case, the formalism of the Hilbert school of mathematics—to serve his own antipositivist, Platonist ends. Far removed, also, from the positivist creed was Einstein's masterpiece, general relativity, which went beyond the special theory of relativity by providing an account of gravity. Further removed still was his lifelong opposition to the positivistic Copenhagen interpretation of quantum mechanics championed by Heisenberg and Bohr. In Einstein, the positivists would soon discover, they had acquired not a friend but an enemy.

The same lesson was learned, the hard way, by David Hilbert, a towering figure in mathematics, who, inspired by Einstein, had formulated the equations of general relativity five days before Einstein himself succeeded, a situation which led, unsurprisingly, to some uncomfortable moments in their relationship. The positivistic creed—by its own nature as opposed to the spirit of mathematics as to philosophy—had in the course of time found a home in mathematics as well. As the positivists would have it, the hierarchy of transfinite numbers discovered by Georg Cantor, a surprising consequence of his theory of sets, was cast into disrepute for bearing the stain of Platonism, for pointing to infinite horizons beyond the frame of the natural realm. The great Hilbert, however, defended Cantor's set theory, proclaiming, "No one shall expel us from the paradise that Cantor has created," and calling it "one of the supreme achievements of purely intellectual human activity."

Cantor's paradise was a lush tropical domain of infinities that he claimed to have encountered at the very heart of mathematics. The importance of a sound theory of infinity was lost on neither mathematicians nor physicists. Mathematics, a tool indispensable to physicists, had been undergoing a gradual development of increased rigor and clarification of foundations, a process that came to fruition in the second half of the nineteenth century and the first years of the twentieth. Infinity played an essential role. Once and for all, it seemed, a firm foundation had been laid for the calculus invented by Newton and Leibniz, in which so-called infinitesimals (infinitely small quantities) enjoyed an ambiguous twilight existence between finitude and infinity. Weierstrass, Cauchy, Cantor and others developed the modern theory of limits of infinite sequences, which for the first time made rigorous sense of Newtonian concepts like "point" and "instantaneous velocity." Further, Cantor, Frege, Dedekind and others put forward a convincing theory of real numbers—rational numbers as infinite sequences of natural numbers, and irrational numbers as infinite sequences of rational numbers—which was crucial, since the physical continuum of space and time could be fully described only by the real numbers. (Frege also advanced an account of the natural numbers in terms of infinite aggregates of concepts, but this fell on deaf ears.) All of this required, however, a comprehensive mathematical theory of sequences, or more generally groupings, sets or classes of numbers, as well as a mathematical account of infinity. Cantor, in a single bold move, developed precisely what was needed, a set theory that provided a rigorous account of infinite sets.

His first discovery was that the requisite infinity had to be "actual," which went against a two-thousand-year tradition in mathematics, from Aristotle to Gauss, which held that infinity is merely "potential." Before Cantor, it was axiomatic that infinity was not to be considered a definite number. To say, for example, that the natural numbers are infinite in number was taken to mean not that there is an actual number, infinity, that numbers the set of natural numbers, but rather that the set of natural numbers goes on forever, and that the most that one can say is that no natural number is big enough to number the entire set. Cantor, in contrast, produced a powerful argument for the thesis that there is an actual number, which he called $\aleph_0$ (aleph null), that numbers the set of natural numbers. Naturally, he emphasized a fact that we can put as follows: The number that numbers the natural numbers cannot itself be a natural number. It must be an unnatural, or supranatural, or (as Cantor characterized it) *transfinite* number. The king cannot arise from the class of peasants. What established the significance of such a transfinite number was Cantor's second great discovery, a proof that $\aleph_0$ is not the only transfinite number but only the first, the smallest. His proof is one of the wonders of the world, like the hanging gardens of Babylon or the pyramids of Egypt.

Cantor, along with Frege, had introduced a rigorous and mathematically fruitful definition of when two sets are the same size, namely, when their elements can be put into one-to-one correspondence. In his proof, he assumed, by way of contradiction, that there exists a one-to-one pairing between the natural numbers and the real numbers. Using this pairing, he succeeded in constructing, via what came to be called a "diagonal argument," a real number that differed from every other real number in the supposedly complete list. Any attempt to pair off the natural numbers with the real numbers will fail. It follows that there are *more* real numbers than natural numbers, even though there are infinitely many natural numbers and infinitely many real numbers. Since it can be shown that the number of real numbers is the same as the number of subsets of the natural numbers, namely, $2^{\aleph_0}$, it follows that $2^{\aleph_0} > \aleph_0$. By generalizing his argument, Cantor was able to show that the power set of *any* set is always larger than the original set, and therefore, for *any* number, including a transfinite one, there will always exist another that is strictly greater. Thus, not only is infinity an actual number, but there is an infinity of infinities.

Needless to say, infinity is not accessible to the five senses, and an infinity of infinities was clearly too much for any self-respecting empirically minded positivist to bear. Yet as a mathematician as well as a physicist, Hilbert realized that mathematics could not do without Cantor's new foundation for the theory of sets and infinities, and thereby for real numbers and the calculus, indispensable for physics. He therefore took it upon himself to make certain, as he put it, that no one would ever be driven from Cantor's paradise. But a defense was needed for the set theory that Cantor had constructed.

From its inception, set theory was haunted by paradoxes and conundrums, which served only to make the skeptics more skeptical. For one, as Cantor himself proved, the very "universe" of set theory could not itself be a set. There is, provably, no universal set, no set of all sets. The reason, in a nutshell, is that if there were such a set, its number would have to be larger than any other transfinite number. As we have already seen, however, Cantor proved that for any transfinite number, there is always a larger one. Set theory, conceived of as a foundational science, was unable to account for itself. It failed even to tell us how many sets there are.

The coup de grâce, however, to unreconstructed, or "naïve," set theory was the paradox that Bertrand Russell discovered in its very foundations. Every concept, every property, it was thought, determines the set of things that have this property. The property of being a horse determines the set of horses; the property of being a prime number, the set of prime numbers; and the property of being a small set, the set of small sets. At worst, the set or class of things determined by a given property is empty. As things turned out, however, this was by no means the worst that could happen. Russell, annoyingly, asked us to consider the property of being a set that is not a member of itself. The set of small sets, for example, is not a member of itself (since it is clearly not a small set), whereas the set of big sets surely is. Russell was able to show, however, that there could not be such a thing as the set of all sets that are not members of themselves. If there were such a set, it would have to be and also not be a member of itself. It follows that it is not true that every property determines the set of things that have that property. But then, which properties do determine sets, and more generally, exactly which sets actually exist?

Russell's paradox was disarmingly simple. It left mathematicians breathless. How, one wonders, did Russell ever come up with his dangerous idea? Historical research has revealed that he invented his paradox in the course of trying to refute Cantor's proof, rehearsed above, that there are more real numbers than natural numbers. His arrow missed Cantor but struck Frege squarely in the chest, toppling his formal development of set theory and shattering his life's work. We still have the polite and lethal letter Russell sent to Frege in 1902: "Dear Colleague, . . . I find myself in agreement with you in all essentials. . . . I find in your work discussions and distinctions . . . one seeks in vain in the works of other logicians. There is just one point where I have encountered a difficulty. . . ." Russell's paradox threw not just set theory but mathematics itself into a crisis, the third great crisis in the history of mathematics. The first had taken place when the Pythagorean theorem revealed to the ancient Greeks the existence of irrational numbers, those that cannot be expressed as a ratio of two natural numbers. The second came when Newton and Leibniz founded the infinitesimal calculus on the basis of infinitesimal numbers, which were supposed somehow to be simultaneously nonzero and yet count for nothing. The crises had a common cause: mathematicians found themselves confronted with a paradoxical new kind of number. If a way could not be found to incorporate this new entity into their thinking, they were faced with the prospect of seeing their edifice crumble.

"The sole possible foundations of arithmetic seem to vanish," Frege wrote, when confronted with Russell's paradox.

With the third crisis, the positivists' star had risen. Mathematics itself, by its very nature as an a priori, rationalistic science, had always been a thorn in the side of empiricists. But now, with Cantor, mathematics had seemingly overreached itself. It had tried to fly too high in the thin air of infinity and was in danger of crashing down on the solid earth below, the empirical soil on which natural science is based. For mathematicians like Hilbert who were also, in spirit, positivists, this engendered a crisis of divided loyalties. A way must be found somehow to preserve Cantor's mathematical paradise. The answer, for Hilbert, was to reconstruct mathematics itself along the lines of positivism. The formal proof of the mathematician would serve as an analogue of the measuring apparatus of the empirical scientist. Formal mathematical proofs—which can be written down on a blackboard and perceived with the senses—are, no less than the instruments of the physicist, things you can actually "get your hands on." Hilbert, then, was the Moses who would lead mathematicians through the desert of positivism back to Cantor's paradise. He would preserve the letter if not the spirit of Cantor's theory of infinite sets, in a manner that satisfied the strict epistemological requirements of positivism.

Some years later, Kurt Gödel would describe positivism as but one element, along with skepticism and materialism, in what he called the dominant "leftward" worldview. The "rightward" view, in contrast, was characterized by spiritualism, idealism and theology (or metaphysics). "The development of philosophy since the Renaissance," said Gödel, "has by and large gone from right to left," reaching a high water mark in the positivistic Copenhagen interpretation of quantum mechanics. "Mathematics," however, "by its nature as an a priori science, has in and of itself an inclination toward the right." But mathematics could not escape the zeitgeist, and so "around the turn of the century its hour struck: in particular, it was the antinomies of set theory, contradictions whose significance was exaggerated by skeptics and empiricists and which were employed as a pretext for a leftward upheaval."

The result of this upheaval was Hilbert's brainchild, the mathematical program known as "formalism," in which the intuitive notion of mathematical *truth* was to be replaced by the formula game of *proof* from a list of axioms according to a set of rules of derivation. "Thus came into being," wrote Gödel, "that curious hermaphroditic thing that Hilbert's formalism represents, which sought to do justice both to the *Zeitgeist* and to the nature of mathematics." Justice was to be done to the spirit of the time by refusing to acknowledge the fundamental axioms of mathematics as in any sense true, and by declaring that the inferences to be drawn from these axioms are, in Gödel's words, to be "construed as a mere game with symbols according to certain rules, likewise not [supported] by insight." The "implicit definitions" of such a formula game did not aim at a true account of the fundamental entities, such as points, lines, or numbers; rather, a point, line, or number was "by definition" simply anything that satisfied the axioms. To this conception the positivist attached himself like glue, and refused to consider the possibility of an alternative picture according to which there exists a priori insight into an objective mathematical world. But such stubbornness on the part of self-described empiricists, Gödel remarked, was really no more than "an a priorism with the sign reversed."

Just as Mach's supposedly hard-headed empiricism left him adrift on an ocean of purely mental phenomena, so the formalist rejection of the very idea of mathematical truth turned mathematics into a purely mental construct, a mere game with formulas, with no intrinsic connection to the physical world. As Schlick wrote, "A system of truths created with the aid of implicit definitions does not at any point rest on the ground of reality." This had the effect of making what Eugene Wigner, decades later, would describe as "the unreasonable effectiveness of mathematics in the physical sciences" even more unreasonable.

Hilbert's formalism, so beloved of strict empiricists, was a crowning achievement of positivism, the most articulate and well-developed attempt to put the positivist's money where his mouth is, an attempt to prove once and for all that the apparent Platonism of mathematics—the very foundation, from Plato on, of all philosophical Platonisms—could be sidestepped. Gödel encountered this achievement of Hilbert's in Vienna, the city in which he lived and worked, the home of the Vienna Circle. Indeed, his own dissertation advisor at the University of Vienna, the mathematician Hans Hahn, was a positivist who concluded a famous essay, "The Crisis in Intuition," with these words: "It is not true, as Kant urged, that intuition is a pure a priori means of knowledge, but rather that it is a force of habit rooted in psychological inertia." And the patron saint, as we have seen, of the Vienna Circle was Wittgenstein, its bible, the *Tractatus,* which declared that the only real facts are of the empirical variety. "The essence of this view," Gödel noted, "is that there exists no such thing as a mathematical fact." Mathematics, for Wittgenstein, consists merely of the transformations of formulas to obtain mathematical identities. In particular, it is not derived from mathematical facts, whereas physics is derived from facts about the physical world.

Between Hilbert and Wittgenstein, it seemed, the positivists had finally laid to rest the ghost of mathematics, which had seemed to resist incorporation into their fold. The ghost, however, was hardly dead, as Gödel would demonstrate from within the heart of Wittgenstein country, Vienna. His incompleteness theorem was a grenade aimed at Hilbert that landed in the very laps of the positivists. Vienna's circle could not after all be completed.

4 | A Spy in the House of Logic

Every spy's life has ended in ignominious death.

ANAÏS NIN

Russell's Paradox, communicated to Frege in 1902, engendered widespread fear in mathematical circles. As Gödel put it, in retrospect, "[Russell] brought to light the amazing fact that our logical intuitions . . . are self-contradictory." On pain of contradiction, it could no longer be assumed that every property determines the class of things that have that property. It could still be trusted that the property of redness sufficed to determine the set of all red things, but such confidence was no longer justified for every property. The concept of class or set, in particular, which had assumed increasing importance in mathematics and logic and which seemed intuitively clear, turned out to be so poorly understood that in Frege's epoch-making formulation of modern logic it led to a straightforward contradiction. The concern was not a communist under every bed but a paradox asleep on every mathematical bedspread.

There was no place to hide. Frege had built a contradiction into the very foundation stone of the mighty edifice of logic. Cantor, the creator of modern set theory, had also constructed, if not contradictions, then at least paradoxes aplenty in his dizzying hierarchy of transfinite numbers. Even Euclidean geometry, the ancestor of all models of mathematical rigor and certainty, the soft pillow for two thousand years of sweet mathematical dreams, had become suspect due to the recent demonstration of the logical consistency of non-Euclidean geometries. If Euclidean geometry could be rejected without contradiction as the truth about space (as arithmetic is the truth about natural numbers), wherein did its truth lie? Everywhere, intuition was under suspicion. If the house of logic was not itself secure, where else could safety be found? Something had to be done.

In Göttingen, the dean of mathematicians, David Hilbert, declared, "Where else would reliability and truth be found if even mathematical thinking fails?" Cantor's paradise, in particular, had to be shored up. "The definitive clarification of the nature of the infinite," said Hilbert, "has become necessary, not merely for the special interests of the individual sciences, but rather for the honor of human understanding itself." Having spent the first years of the century in wide-ranging research, he turned his attention in the century's second decade to the crises infecting the foundations of his mighty edifice, devoting his considerable resources to a solution.

In Vienna, Kurt Gödel was becoming part of the problem, though in 1930 he did not yet know it. Even Hans Hahn, his thesis advisor and a true believer in the positivist credo, discovered too late the full extent of Gödel's heresy. "As concerns the world," Hahn declared, "the only possible standpoint seems to me to be the empiricist one. . . . Knowledge concerning reality can in no way be obtained through pure thought." Yet mathematics and logic, the tools of scientific empiricism, were not cooperating. As Hahn put it, "A very simple fact now seems to stand in the way of realizing this empiricist standpoint, namely, the existence of logic and mathematics." These words were uttered at the very conference at which Gödel would rise to defeat the last best hope for the positivists to incorporate mathematics within their religion of ultraempiricism. And he would do so by exploiting Hilbert's own weapon of choice: the formalism of mathematical logic. Gödel, in short, would destroy from within. This is the reason they shoot spies.

In the case of Gödel, the formalisms he employed, though in themselves acceptable to the positivists, were Janus-faced by design. On one hand, they were irrefutable mathematical theorems. On the other, their most natural, indeed irresistible philosophical implications would undermine the very spirit of positivism. It was a case of using the letter of a false doctrine to overthrow its spirit. And this was a specialty of Gödel's, as it was of his future friend Einstein: constructing mathematical formalisms pregnant with philosophical meaning. It was a talent guaranteed to arouse ire on both sides: the friends of formalism would be hard pressed to reject philosophical implications derived from inescapable premises, while the enemies of formalism would be forced to admit that something philosophically significant could after all be achieved within the narrow confines of the formal.

The Leitmotif of the Twentieth Century

> *who pays any attention*
> *to the syntax of things*
> *will never wholly kiss you.*
>
> e.e. cummings

Gödel's incompleteness theorem of 1931 began innocently, as an attempt not to refute but to fulfill Hilbert's program. Hilbert's idea was to safeguard mathematics from hidden contradictions by replacing the intuitive mathematics of each mathematical domain with a system of axioms written in a pure formula language that, although having a standard semantic interpretation, could be manipulated according to the mechanical rules of pure syntax (much like a computer program of today). Hilbert's program consisted in finding a system of primitive formulas called axioms from which, according to fixed rules of proof—rules of syntax—one could derive all the theorems of the given mathematical domain. Two features of such a formal system were crucial: consistency and completeness. As a prophylactic against unwelcome surprises, a formal system had to be consistent: two theorems that contradict each other should not be able to be derived from the axioms. And the system should be complete, in the sense that all true statements expressible within the system (under a suitable interpretation) should be derivable from the axioms. To prevent circularity, the system in which consistency is to be proved must not itself employ any mathematically suspect or controversial procedures that could render its own consistency suspect. It must be, to use Hilbert's invented term, not exactly finite but rather "finitary," in the sense that its proofs must be in principle surveyable by sense experience and must not at any point appeal to an abstract, completed infinity of the kind proposed by Cantor.

Hilbert's formalism was just one example—the most rigorous, mathematical one—of the spirit that lies at the very heart of the twentieth century. At its core it involves the dominance of form over content, syntax over semantics, proof over truth. It is no surprise that the principal embodiment of a formal system, the computer, a pure syntax machine, would become the century's dominant mechanical device. But the computer was still just one element of the zeitgeist. In art, science, philosophy, mathematics, music, architecture and linguistics, formalism in its most general sense became the dominant theme. In painting, for instance, Cezanne's realism was a hidden case of the free play of geometrical forms, a fact his subjects came increasingly to appreciate as they realized that the geometrical constraints of his canvases dominated any attempt to capture the shape or spirit of those who sat before him. This paved the way for the explicit rule of the play of free forms by the cubists, led by Gris, Braque, and Picasso. The Cezanne of music was Brahms, whose post-Romantic chromaticism hid the dominance of pure logical structure at its core. Wittgenstein, in whose Vienna home Brahms performed, put the matter darkly: In Brahms, he said, "I can begin to hear the sound of machinery." This hidden formalism, this logical machinery, was not lost on Brahms's admirer Schoenberg, who would soon champion the freely constructed forms of serial music, the most explicitly conventional, mathematical method ever undertaken in music. The principal standard-bearer of Schoenberg's piano music, Glenn Gould, would speak of producing a structural "x-ray" of the score in his performances. As the high priest of bones without flesh, Gould made it clear that his first god was Bach, whom Schoenberg also worshipped, his gospel, the fugue.

Nor was physics left behind. On the contrary, it was in the vanguard. In special relativity, Einstein had abandoned the Kantian intuitions of space and time for the mathematical formalism of space-time, constrained only by the formal requirement of Lorentz invariance and the physical postulate of the limiting value of the speed of electromagnetic signals. General relativity, the more inclusive theory, would yield an abstract structure governed by yet more general logical constraints. And the cognitive and social sciences would follow physics' lead. Later in the century, Noam Chomsky would re-create linguistics as a structural science modeled on Frege's logic, with syntax explicitly dominant over semantics, while Claude Levi-Strauss would stitch the abstractions of structuralism into the many-colored quilt of anthropology. In mathematics itself, the trend was increasingly toward the reduction of a domain into the structural relationships that obtained between its elements. What mattered, Hilbert insisted in his reconstruction of Euclid's geometry, was not what points, lines and planes are, i.e., the semantics of the fundamental terms, but the logical relationships that obtain between the basic elements, i.e., the syntax of the formal system. As Hilbert put it, a point in Euclidean space could be a beer mug for all he cared, as long as it obeyed the rules of his formal system. This idea of a so-called implicit definition of a concept via its relationships with other concepts was firmly rejected by the father of the formal system, Gottlob Frege. In his account of the natural numbers, Frege objected that his own preliminary contextual or implicit definitions failed to determine what the individual numbers actually are, in particular whether some collection could have the number Julius Caesar attached to it, nor "whether that same familiar conqueror of Gaul is a number or not." This was a remark as crazy as it was beautiful. No one, of course, is likely to confuse Julius Caesar with a number. But that is not, Frege pointed out, due to the power of the implicit definitions.

In the decision to reduce all to syntax, to focus on form alone, could be found the freedom of the creative imagination. There could also be found safety. Because the rules of the formal system are our own creation, we would be able to police them, to examine them as pure signs to see that they did not lead to inconsistency, to contradiction. Consistency, not truth, increasingly became the goal of the formal systems of science, just as authenticity became the battle cry of the systems of ethics or forms of life. Sartre's existentialists, for example, attempted bravely, and perhaps foolishly, to replace conscience with authenticity. The problem, of course, was that Hitler too was authentic. It was the content of his beliefs that was the problem, not their consistency. Since mathematics is the language of formal relationships, it became increasingly clear that the central formalism was that of mathematics itself, and that if this could not be rendered secure from inconsistency, nothing else could. If, then, formalism is the leitmotif of the twentieth century, and Hilbert's mathematical formalism captures the essence of all other formalisms, then Gödel's incompleteness theorem, which dramatically and inescapably refutes Hilbert's program, can well be considered the most significant intellectual accomplishment of the twentieth century.

Gödel took up Hilbert's momentous project by attempting first to see whether one could prove the consistency and completeness of a formal axiom system for mathematical analysis. He began with the task of proving consistency and completeness for the weaker axiom system of arithmetic, or number theory, a subsystem of analysis. Here the conditions were propitious. For thousands of years, geometry had been the paradigm of an axiomatic system, but in the late nineteenth century, Frege, Dedekind and Peano had achieved the same result for arithmetic. They constructed a system of axioms, or postulates—known today, for no good reason, simply as the five Peano postulates—from which it was believed that all truths about the natural numbers could be derived. But derived how? By logic alone, the logic that Frege had invented in his *Begriffsschrift*. Whereas the axioms had, semantically speaking, genuine mathematical content, the rules of inference were a matter of pure syntax, a series of mechanical instructions that could be followed blindly, with no reference to truth or mathematical content, followed, as the logician John Myhill put it, by an imbecile or a computer.

What Gödel discovered, however, was that not only are the Peano postulates in fact incomplete, any system of axioms or postulates (even if infinitely large) from which arithmetic can be derived that satisfies any reasonable mathematical criteria of surveyability by a finite mind is of necessity incomplete. (An infinite mind, like God's, which can grasp all the numbers at once, presumably has no need of axioms.) So the simplest and most basic domain of mathematics, the arithmetic of the natural numbers, the rock on which the grand edifice of mathematics stands, turns out to be, from a formal axiomatic point of view, incomplete, and even worse, incompletable. Indeed, since a computer can prove only theorems based on the axioms its programmer has fed into it—it cannot, as Gödel emphasized, create new axioms on its own—it follows that in principle no computer or fully specified system of computers, even if infinite, will ever capture all the truths of arithmetic (never mind the rest of mathematics). As Gödel put it, "Continued appeals to mathematical intuition are necessary ... for the solution of the problems of finitary number theory."

The mathematical fact of the incompleteness of formal arithmetic, moreover, is accessible not only to us, thinkers with minds and mathematical intuitions; ironically, a computer can be programmed to prove Gödel's theorems, the very theorems that establish the intrinsic limitations of computers. The truths of arithmetic, then, cannot in principle be confined to a formal system. Here is a crucial difference between truth and proof: a mathematical proof, in the sense in which we are discussing it here, is always a proof in, and relative to, a given formal system, whereas truth, as such, is absolute. What Gödel proved is that mathematical truth is not reducible to (formal or mechanical) proof. Syntax cannot supplant semantics. The leitmotif of the twentieth century, it turns out, stands in need of revision. Mechanical rules cannot obviate the need for meaning, and what gives us access to meaning, namely, intuition, cannot be dispensed with even in mathematics, indeed, even in arithmetic. This was the first nail in Hilbert's coffin.

The second nail was not long in coming. Gödel soon proved his second incompleteness theorem, which demonstrated, with yet further irony, that if a given system of axioms for arithmetic were in fact consistent, then it could not be proved consistent by the system itself. Put otherwise, only an inconsistent formal system can prove its own consistency! Von Neumann, the quickest of the quick, having heard Gödel announce his incompleteness results, derived, shortly thereafter, the unprovability of consistency. "I would be very much interested," he wrote Gödel, "to hear your views on this. . . . If you are interested, I will send you the proof details." One can imagine his disappointment when Gödel informed him that the manuscript for the second theorem was already on its way to the editors. It was Von Neumann, however, who argued, against Gödel himself, that the unprovability of consistency, as Gödel had demonstrated it, left no wiggle room for the Hilbert program. Whereas for several years, Gödel was cautious not to prejudge the question of whether Hilbert might discover a finitary proof of consistency to which Gödel's second theorem did not apply, Von Neumann, from the beginning, was confident that this could never happen. Assuming that one rejects Russell's controversial "axiom of reducibility," he said, "one cannot obtain a foundation for classical mathematics via logical means." Von Neumann's striking prescience, however, concerning the full significance of what Gödel had discovered may well have served only to deepen his regret that he had not been the first to make these discoveries. Even the fact that he was one of the fathers of the modern computer and a chief architect of the atomic bomb in Los Alamos did not suffice to assuage his disappointment.

The incompleteness theorems sent a shock wave through the world of mathematics. Hermann Weyl, one of the first permanent members of the mathematical faculty at the Institute for Advanced Study, spoke of the Gödel "debacle," the Gödel "catastrophe." The two-thousand-year-old ideal of axiomatization inaugurated by Euclid—the paradigm of captured rationality—had been shattered, and the blow had been struck, annoyingly, just when Frege and Hilbert had succeeded in perfecting the very idea of a formal system of axioms. Not only the results but the very methods employed in Gödel's theorem were so unexpected that it was years before mathematicians and logicians began to grasp their full significance. Gödel had carried even further than Hilbert the idea of treating formal systems of mathematics as mathematical objects in their own right, which resulted in conclusions exactly opposite to what Hilbert had intended. (This is an instance of what Hegel called the "cunning of history," whereby history itself contrives, somehow, to subvert the intentions of its most dramatic actors.)

In the case of Gödel's theorem, as with his later writings on relativity, the difficulty in taking the true measure of its significance was due not just to the mathematical but to the philosophical hurdles that had to be overcome. In the next few paragraphs, we will take a first-hand look at some of the details of Gödel's mysterious and beautiful proof. If the going gets tough, the tough may either get going or they may, without loss, simply skim lightly over the details to get the gist of Gödel's argument, or they may even take off, fly over and pick up the thread at the clearing after the forest. Do not, in any case, be intimidated; to switch images, you can admire the music without attending to the words. To appreciate Gödel's theorem is your birthright; let no one, including the mathematical police, deprive you of what you have a right to enjoy.

Bear in mind also what Gödel proved and what he did not. He did not discover some deep and mysterious mathematical proposition that no formal system was powerful enough to count among its theorems. That would have demonstrated the existence of an absolutely unprovable mathematical proposition, something that Gödel, like Hilbert, was deeply skeptical of. Rather, what he showed is that in any particular formal system of sufficient strength, given the limitations imposed on such a system insofar as it is truly formal, there would always be some formula which, while intuitively true, could not be proved in or relative to that system. And the same holds for its negation. But the formula would be a perfectly ordinary, though complex, mathematical proposition, which nevertheless, because of its form, slipped through the net of the given formal system. That very formula, however, could always be proved in a more inclusive formal system; only that new formal system, in turn, would be unable to prove some new formula, which was nevertheless intuitively true. And so on. There was, then, no "supervirus" that affected all formal systems. Instead, for each particular formal system, there would be some perfectly ordinary bug or virus that rendered that system incomplete.

Triple Fugue: Intuitive Mathematics, Formal Mathematics, and Metamathematics

Gödel's beautiful fugue was constructed from three distinguishable mathematical languages or theories. The beauty was to be found in the pattern of relationships woven from the three parts. To begin with, there was intuitive arithmetic, the arithmetic found in mathematical textbooks written in the language of ordinary mathematics. Call this language or theory IA (for intuitive arithmetic). The propositions of IA are sentences with content: they express truths or falsehoods about numbers. Next there was a formal deductive system for arithmetic—in Gödel's proof, a system of pure syntax put forward by Bertrand Russell, modeled on the original by Frege—with a specified set of axioms and explicit rules of deduction that determined which formulas were theorems. Call this system FA (for formal arithmetic). The "sentences" of FA are simply formulas without semantic content. In themselves, they are neither true nor false. They are, however, either provable from the axioms of FA or not. If provable, they are called theorems. FA, however, is so designed that we can give it an interpretation, a semantics, under which it can be read as corresponding to IA. That is, FA is designed to mirror IA, so that if all goes well, there will be an exact one-to-one correspondence between the numerals in FA and the numbers in IA, and a similar correspondence between the true sentences of IA and the theorems of FA. Put succinctly, FA is designed to represent IA.

The third language or theory is the metatheory of formal arithmetic, the framework in which the syntactic rules, the proof theory, of FA is spelled out. Call this language MFA (for the metatheory of formal arithmetic). If FA is the machine, MFA is the owner's manual that specifies how the machine works. Like IA, it consists of meaningful sentences that have truth values. MFA specifies which formulas of FA are "well-formed formulas," meaning that they satisfy the official rules of formula construction. Crucially, it also specifies what it means to be a proof in FA.

Gödel's insight was to see that FA could be used to represent not only IA—to the extent to which this is possible—but also MFA. He proved the latter by showing how MFA could be represented in IA, via a revolutionary device known today as the "arithmetization of metamathematics." But then if FA can represent IA, it can also, via the interpretation of MFA in IA, represent MFA. That is, FA can represent its own metatheory. The trick, then, was to construct a formula of FA that would have two simultaneous meanings in two languages, MFA and IA. Gödel was able to exhibit just such a formula and to prove that it was simultaneously unprovable in FA and, intuitively, true in IA and MFA. This would be a formula provably unprovable in FA, and yet expressing a true proposition in IA about the natural numbers as well as a true proposition in MFA about its own unprovability.

Nothing like this had ever been seen before. Gödel had skirted around the deadly liar's paradox, substituting for it an unproblematic unprovability paradox (which was not really a paradox at all); established the possibility and harmlessness of self-reference; demonstrated representability relationships among three distinct languages; arithmetized the syntax of one of those languages; and finally, exhibited a formula of one language that was provably unprovable and simultaneously true. This was logic, it was mathematics, but it didn't look like logic or mathematics. It looked more like Kafka. Indeed, when the mathematician Paul Cohen, a Fields medalist who proved the independence of the continuum hypothesis, first encountered Gödel's theorem, he was skeptical, remarking that it seemed more like philosophy than mathematics. After discussing the theorem with the logician Stephen Kleene, however, his doubts evaporated. Still, "I was rather depressed," he commented later, "when I realized Gödel was right."

Reusable Numbers: The Arithmetization of Syntax

Gödel's first task was the easy one, showing that FA could be used—at least a fortiori—to represent IA. The hard work had already been done by Frege, Dedekind and Peano in establishing the Peano postulates, finding the crucial definitions of number-theoretic concepts needed to employ these postulates to represent the facts of number theory, and constructing a system of strict logical rules that would specify precisely which inferences were permitted in FA. The second task, showing that FA could stand in for MFA, was the real challenge. Not only its implementation but the very suspicion that it could be done required a stroke of genius.

In effect, Gödel was borrowing a leaf from Descartes's book. Descartes, by assigning numbers to figures in geometrical space through what are now known as Cartesian coordinates, was able, as we would now say, to "arithmetize geometry." In this system of so-called analytic geometry, statements about geometrical figures are translated into statements about numbers, and the powerful rules of manipulation of numbers, in turn, can be exploited to make discoveries about geometry. Gödel's insight was to realize that the elements of the formal system FA—primitive signs, sequences of these signs (such as formulas), and sequences of formulas (including proofs)—could also be assigned a numerical representation. Through a system now known as Gödel numbering, Gödel assigned unique natural numbers systematically to every primitive symbol of FA, and then showed how to construct, again uniquely, the number assigned to a sequence of primitive symbols, i.e., a formula, and to a sequence of formulas, including a sequence that constituted a proof. To ensure uniqueness in the system of numbering, Gödel relied on well-known facts about natural numbers such as the fundamental theorem of arithmetic: every positive integer can be resolved uniquely into a product of prime numbers. This arithmetization of syntax established, in effect, a language translation scheme between symbols of FA and natural numbers, so that a statement about the syntax of FA could be translated, via the rule book of Gödel numbering, into a statement about natural numbers, just as in the Cartesian system, a fact about figures in the geometrical plane could be translated into one about real numbers.

So far, the system of Gödel numbering as we have described it only sets up a correspondence between numbers and symbols, sequences of symbols, and sequences of sequences of symbols of FA. But there is more to the syntax of FA than this. There is also the question whether a given formula of FA is a well-formed formula, and, crucially, the question whether a sequence of formulas constitutes a proof. What Gödel proved is that all the crucial functions needed to describe the complete syntax of FA, including being a well-formed formula and being a proof of FA, corresponded to certain recursive functions in IA. A recursive function is one that, intuitively speaking, can be mechanically computed. This kind of function can also be characterized strictly mathematically, and this Gödel proceeded to do. An example of a so-called primitive recursive function, the "+ function," also known as the "addition function," will illustrate what is meant by recursivity. Let us call the number that comes right after a natural number x the successor of x, or $s(x)$. The "+" function, then, is given by two rules:

(a) $x + 0 = x$;
(b) $x + s(y) = s(x + y)$. (This can be read aloud as "x plus the successor of y equals the successor of x-plus-y.")

The successor of x, namely, $s(x)$, can be defined as $x + 1$. This kind of recursive definition can be used to compute mechanically, by a kind of "bootstrapping," the sum of any two natural numbers, since every natural number is either 0 or the successor of some other natural number.

Recursive definitions were studied by Dedekind, Peano, Skolem and others, and recursive functions had been used implicitly throughout the history of mathematics, but the first to elaborate a precise and forceful account of such functions was Gödel, who cited his young French colleague Jacques Herbrand as having influenced his understanding of these ideas. Herbrand had written to Gödel on hearing of his incompleteness results from Von Neumann. Gödel wrote a detailed and deeply respectful response, at the end of which he suggested that in the future they correspond each in his mother tongue. (Gödel was very good at languages.) He never received a reply. What he did receive was a touching letter from Herbrand's father informing him that the reason for his son's silence was that he had fallen to his death while climbing in the Alps. Jacques Herbrand had been just twenty-three years old.

Gödel demonstrated, then, that the fundamental concepts of MFA, in which was found the metamathematics, or proof theory, of FA, corresponded to certain recursive functions in IA. In particular, the function Bew(x, y), i.e., x is a proof of y (from the German for proof, Beweis), when coded into natural numbers, yields a recursive function. This was important because in proving that FA can represent IA, Gödel had already shown that any recursive function contained in IA could be represented in FA. Specifically, if there was a truth about a recursive function in IA, there would be a corresponding formula that was a theorem of FA. Once he had demonstrated that the basic functions in MFA when coded into natural numbers yield recursive functions, he could conclude that MFA, just like IA, could be represented in FA. Further, via Gödel numbering he had already arithmetized the syntax of FA, so a fact about the syntax of FA would correspond to a fact in IA about the natural numbers. Gödel had shown, then, that the theorems of FA could represent simultaneously the arithmetic truths of IA as well as, via Gödel numbering, the syntactic truths of MFA. That is, a given theorem of FA would represent a mathematical truth in IA that would itself, via Gödel numbering, represent a syntactic truth about FA. Gödel had succeeded in proving, then, that FA, though in itself a system of formal, meaningless signs, could be "double stuffed" with meaning, i.e., assigned meanings that ensured that it could be used to represent, simultaneously, number theory and the syntax or proof theory of FA itself.

In other words, FA could speak about itself via the natural numbers. The natural numbers, Gödel had shown, were "reusable," in the spirit of Descartes: they could be used as elements of arithmetic and at the same time as representatives of the syntax or proof theory of formalized arithmetic.

I Cannot Be Proved

With these elements in place, the coordination of the three languages IA, FA, and MFA and the implementation of Gödel numbering, Gödel made his move. He constructed a formula of FA—known, familiarly, today, as the Gödel formula, or G—whose interpretation in IA was a statement, either true or false, about the natural numbers. But G also had, via Gödel numbering and the arithmetization of the syntax of FA, another meaning. On this interpretation it asserted a proof-theoretic fact about a certain formula of FA, via the Gödel number assigned to that formula, to the effect that the formula with that Gödel number g was unprovable in FA. But Gödel had set it up that the formula whose Gödel number was g was none other than G! In effect, then, on one interpretation, what G stated was, "I cannot be proved." The question then was, Is G provable in FA? (Equivalently, is G a theorem of FA?)

Clearly, this question was a close cousin of the ancient liar paradox in which a sentence S says of itself, "I am not true," whereupon one asks, Is S true? But the liar paradox really is a paradox, or more precisely an antinomy, insofar as the natural answer to the question, "Is S true?" seems to imply both that S is true and that it is not. What Gödel realized is that the question posed about G, which concerned not its truth but its provability, did not lead to paradox or antinomy. The Gödel formula, after all, is simply a formula in the formal system, and as such it was a clear-cut, unproblematic question whether or not G was provable, i.e., was a theorem. Similarly, interpreted as a statement about numbers via the coordination of FA with IA, G was a statement not about itself but about natural numbers, and as such it said something unproblematically either true or false, but not both, about the natural numbers.

So is G a theorem of FA? We assume first that FA is consistent. If it were not consistent, it would obviously be useless for mathematics, and the game would be over. Assume next that G is provable in FA. That means that some sequence of formulas S of FA is a proof of G. Using the abbreviation we introduced earlier, that means that Bew(S, G). But we saw earlier that Gödel had shown that the function, Bew(x, y), when coded into natural numbers, yields a recursive function and is thus representable in FA. The representation occurs via the arithmetization of the syntax of FA, so corresponding to a given syntactical truth Bew(x, y) of MFA, there is an arithmetical truth *Bew(x, y)* of IA that corresponds to a formula **Bew(x, y)** in FA that can be interpreted as saying that the sequence of formulas with Gödel number x is a proof of the formula with Gödel number y, and this formula, **Bew(x, y)**, is a theorem of FA. Thus if G is provable in FA for some s, then **Bew(s, g)** is a theorem of FA (where **s** is the numeral in FA for s, the Gödel number of the sequence of formulas S that constitutes the proof of G, and **g** is the numeral in FA for g, the Gödel number of G). But as we saw at the beginning, what G states is that formula number g (i.e., G) is unprovable, that is, for any sequence x, it is not the case that Bew(x, g). Using a little logic, however, it follows from Bew(s, g) that there exists a sequence of formulas x such that Bew(x, g), hence, using a bit more logic, that it is not the case that for any sequence x it is not the case that Bew(x, g). That is, if something is a proof of formula number g, it can't be that there is no proof of g. But since G can be interpreted as saying that there is no proof of g, this last conclusion is evidently equivalent to not-G, the negation of G, and since it was derived, logically, from **Bew(s, g)**, which, by hypothesis, is a theorem, it too is a theorem. That is, not-G is also a theorem. The assumption that G is provable, then, leads to the conclusion that both G and not-G are theorems, i.e., that both are provable in FA. Hence FA is inconsistent. But the assumption was that FA is consistent. *If FA is consistent, G is not provable.* That is the first part of Gödel's theorem.

How about not-G? Gödel assumed for this half of his theorem that FA is not just consistent but "omega-consistent." (Omega is the traditional symbol for the natural numbers.) If a system is omega-consistent, the following cannot happen: some statement F( ) is provable individually of each natural number, i.e., F(0), F(1), F(2), etc., but it cannot be proved that for all x, the statement F(x) is true. Clearly, if FA is to represent all the natural numbers via its coordination with IA, omega-consistency if nonnegotiable. If not-G is a theorem, then, of FA, given what G says, it follows that the following is provable: it is not the case that every sequence of formulas number x does not constitute a proof of formula number g. But we have already learned that formula number g, i.e., G, is not provable in FA. Thus, for every individual sequence of formulas x, not-Bew(x, g) holds. Given the representability of the function Bew(x, y) in FA, it follows that for every natural number x, it is provable in FA that not-Bew(x, g). Since we are assuming that FA is omega-consistent, it follows that it cannot also be provable that it is not the case that for all sequences of formulas x, not-Bew(x, g) holds. But the latter is precisely what we assumed when we agreed for the sake of argument that not-G is a theorem. *If FA is omega-consistent, not-G is also not provable.* This is the second half of Gödel's theorem. Barkley Rosser later showed that one can construct an unprovable statement assuming only consistency, not omega-consistency.

But Gödel wasn't done yet. He was able to prove, quickly, his second incompleteness theorem, also on the assumption that FA is consistent. Using techniques similar to his proof of the first theorem, Gödel proved easily that if FA is consistent, it cannot prove that it is consistent. Not only was truth not fully representable in a formal theory, consistency, too, could not be formally represented. The Hilbert program had suffered a fatal blow. Gödel had proved that there was in principle no method by which a mathematician, regarding his theories simply as uninterpreted formula games, could prove them free from hidden inconsistency. There simply was no such thing as a magic shield that would resolve all a mathematician's fears of an assault from some unsuspected inconsistency. (Similarly, as a direct consequence of Gödel's incompleteness theorems, there can never be a foolproof antivirus computer program that we can be certain will not alter the program being protected but that will detect the presence of any other program that is attempting to alter the protected program.)

Yet lest we be tempted to complain that Gödel's theorem brought only bad news for our cherished computers, we should recall that the very existence of the modern computer, which is after all no more than a sophisticated calculation or deduction machine, is a direct consequence of the isolation and clarification by Gödel and Herbrand of those recursive mathematical functions that are at the heart of the incompleteness theorem. It was Alan Turing, who developed his idea of what came to be called a Turing machine on the basis of the concept of recursive functions, who became the modern computer's most direct parent. What is crucial to a computer, after all, is that the instructions that are programmed into it can be followed mechanically, syntactically, without recourse to meaning, and in a manner that is "iterative" or "bootstrapping," so that one instruction leads directly to another. This is the heart of the idea of a Turing machine.

These same recursive functions served ultimately to render Gödel's theorem a yet more decisive refutation of Hilbert's program. For the mathematical characterization of a function as recursive, developed by Gödel and Herbrand, was closely followed by a similar characterization, by Turing, of Turing computable functions, and a likeminded characterization of a class of mathematical functions introduced by the American logician Alonzo Church in his calculus of lambda conversion. Soon, Church's students Kleene and Rosser (Turing too, it should be noted, was a sometime student) proved that these three classes of mathematical functions are in a strict sense equivalent. This prompted Church to propound what is today called Church's thesis, which states that this class of mathematical functions, the recursive functions, corresponds exactly to the intuitively characterized class of functions that are mechanically or effectively calculable.

Church's thesis is today widely accepted, and it gives Gödel's theorems yet greater force in two respects. First, it made definitive the characterization of a formal system like FA as one whose syntactic rules must be mechanically specifiable, and thus specifiable by recursive functions. Since Gödel's proof clearly could be extended to any formal system whose syntax was fully characterized by recursive functions, that meant that his incompleteness theorem applied with equal force to any system characterizable as a genuinely formal one. Second, Church's thesis added force to Gödel's second theorem, since it could plausibly be argued that the methods that Hilbert would find acceptably "finitary" were exactly those that could be characterized as mechanically specifiable, and once these were identified with the precisely specified class of recursive mathematical functions, it was an inescapable consequence of the second theorem that it directly refuted Hilbert's program of establishing a finitary proof of consistency for formal arithmetic. Through the work of Church and Turing, then, the full force of Gödel's theorem began to be appreciated.

The Formal and the Intuitive

*God's mercy preserves mathematics
from being drowned in mere technique*

SIMONE WEIL

But it was not always so. The mathematical shot heard round the world began as a whisper. Gödel may have been a spy in the house of logic, a revolutionary, an intellectual bomb thrower, but he was also a citizen of Vienna, the city of coffeehouses. So he first communicated his momentous discoveries to Rudolf Carnap in the Café Reichsrat, shortly before they both left for a conference on the foundations of mathematics in Königsberg (today, Kaliningrad, in which both Kant and Hilbert were born), where Gödel would announce his results publicly. The first to be apprised of what the young man had done, however, seems to have been the first to fail to appreciate it. At the Königsberg conference a few days later, Carnap proceeded, in good positivist style, to recommend consistency once again as the touchstone of formal mathematical theories.

Carnap had been a student of Frege and of the philosopher Edmund Husserl and by 1931 was an established logician. If Frege can with justice be called the father of modern "analytical" philosophy, Husserl can be considered the father of the "continental" branch. The two reviewed each other's books and corresponded, but their followers assembled into armed camps that exchanged intellectual gunfire. The influence Husserl came to have on Gödel late in his life was an anomaly; to his colleagues, it was an embarrassment. In 1931, however, it was not an embarrassment for Carnap to have studied with Husserl. That such a man as Carnap failed to grasp the force of Gödel's accomplishment is an indication of the break with tradition that Gödel was inaugurating (a phenomenon his future friend Einstein had already encountered with relativity theory). Even a year later, Carnap confided that he still found Gödel's results hard to understand.

Von Neumann alone immediately grasped the force of the discovery. He was present at the meeting as one of the principal speakers, and after this encounter he would become a lifelong friend and admirer of Gödel. But Von Neumann, as was so often the case, was the exception. So little impact did Gödel's announcement have on his immediate listeners—the cream of the foundationalist mathematicians and formal logicians—that when Hans Reichenbach, another prominent member of the Vienna Circle, wrote up an account of the conference for the journal *Erkenntnis*, he did not even mention Gödel. The mathematical community, however, was quick to recover from this slight. Immediately after the conference, when news of the paper Gödel had submitted for publication began to circulate, he was invited by the editors of *Erkenntnis* to add a postscript summarizing his soon-to-be published results.

Often, a mathematical result—like Fermat's last theorem—is believed even though its proof is unexpected, so that the appearance of a genuine proof creates a sensation. But in the case of Gödel's theorem, the result itself was unexpected, so the appearance of a proof—a proof like nothing seen before—was explosive. But fame does not bring with it comprehension, and Gödel's methods as much as his results were so revolutionary that it would be years before the mathematical and philosophical communities could fully digest them.

Paul Bernays, Hilbert's assistant and a great logician in his own right, who would go on to become one of Gödel's closest associates, was perplexed by some of the details of Gödel's proofs and asked for Hilbert's assistance. Hilbert's response was at first merely anger, followed by denial. He became the first, but by no means the last, to propose, in effect, an "anti-Gödel" principle: a formal principle, concocted ad hoc, to be appended to formal mathematics simply in order to block the application of Gödel's theorem. Gödel, usually unflappable even in the face of opposition or incomprehension, was genuinely irritated by this. Hilbert's idea—which he would give up soon enough—was to propose that there be adjoined to his formal systems a new rule of deductive inference that would allow for the employment of infinitely many premises. As Gödel pointed out, however, this proposal violated the very idea of a formal system, an idea Hilbert, following Frege, had been at pains to develop. The proposed cure would kill the patient.

If Hilbert did not come off well in his first response to Gödel's theorem, he displayed charm and fairness elsewhere. Two characteristic incidents bring this out clearly. Before World War I, he stood out by his willingness to take on as a doctoral student Jakob Grommer, who had attended a Talmudic school in Eastern Europe and lacked a gymnasium certificate. In addition, Grummer's hands and feet were deformed, which led the daughter of the rabbi he was to replace to reject him. Hilbert, however, did not reject him, declaring that "if students without the gymnasium diploma will always write such dissertations as Grommer's, it will be necessary to make a law forbidding the taking of the examination for the diploma." The second incident concerns the physicist Max Born, who began his academic career in mathematics. Before his examination, he asked Hilbert for advice, and was asked what the area was in which he was the least prepared. It was ideal theory. To Born's dismay, in the examination that followed, just this was the topic on which Hilbert chose to focus his questions. His explanation to Born? "I was just interested to find out what you know about things about which you know nothing."

Hilbert soon recovered from the shock of Gödel's discovery and proceeded to incorporate and develop it in his and Bernays's new textbook on mathematical logic. The same cannot be said, however, for Ernst Zermelo, the mathematician who had inaugurated the axiomatic development of set theory after Russell's paradox had demonstrated that the naïve set theory developed by Cantor lacked a coherent philosophical foundation. Even today, the axioms of Zermelo-Fraenkel set theory are the most widely used and accepted in the field. Yet Zermelo, from beginning to end, was unable to understand or accept Gödel's results. He became their principal mathematical opponent. (Wittgenstein bears the honor of being their chief philosophical detractor.)

Zermelo's difficulties were understandable. Gödel's theorems traded on crucial distinctions such as truth versus proof, semantics versus syntax, and completeness versus formal consistency, distinctions that, though in the air, became fully clarified for the first time only after Gödel's proofs had appeared. It was not that Hilbert, the founder of formalism, distinguished carefully between truth and proof and simply opted for the latter. Rather, as Gödel himself put the matter years later, "formalists considered formal demonstrability to be an analysis of the concept of mathematical truth and, therefore, were of course not in a position to distinguish the two." In the realm of mathematics, proof, for the formalist, was indistinguishable from truth, and so any attempt to draw distinctions between them was simply incomprehensible. Zermelo's philosophical framework, in turn, though different from Hilbert's, was so contrary to Gödel's that reconciliation was impossible.

Fate brought the two men together at another mathematical meeting, this time in Bad Elster, a year after the conference at Königsberg. When it was suggested to Zermelo after the talks were over that he meet with Gödel for lunch on a nearby hill, he demurred, complaining first that he "did not like Gödel's looks," then that the supply of food was insufficient, and finally that the climb would defeat him. Zermelo should have trusted his instincts. He was finally talked into meeting with Gödel, but the encounter, though polite, was fruitless. He would soon write to Gödel that he had a discovered a "major gap" in his argument, and a lengthy reply—running to ten handwritten pages—by Gödel did little to disabuse him of his doubts. Having once failed to enlighten Zermelo, Gödel apparently gave it up as a lost cause, declining to respond even when Zermelo published his criticisms. Carnap, when shown Zermelo's letters, agreed that he had "completely misunderstood" Gödel's achievement.

If Zermelo's intransigence was to be expected, Bertrand Russell's ambivalence was not. The coauthor of the monumental *Principia Mathematica*, which provided the actual formal system for Gödel's proof, continued, late in life, to refer to Gödel's results only guardedly. In a letter written in 1963, Russell, while acknowledging the greatness of Gödel's achievement, did not conceal that he remained puzzled by it, asking rhetorically "are we to think that 2 + 2 is not 4, but 4.001?" This suggests that Gödel had purported to have demonstrated a flaw in classical mathematics, which precisely misses the point of Gödel's theorem. Russell knew, however, that he had not yet fully thought this through. He commented, dryly, that he was "glad [I] was no longer working at mathematical logic." Apparently, Gödel was too. In a letter to a colleague he wrote that "Russell evidently misinterprets my result; however, he does so in a very interesting manner. . . . In contradistinction

Wittgenstein, in his posthumous book, advances a completely trivial and uninteresting misinterpretation."

Still, most logicians and mathematicians came to appreciate Gödel's achievements, which eventually assumed their rightful place as part of the new orthodoxy. Hilbert's program was largely abandoned, in company with other mathematical misadventures like squaring the circle and proving Euclid's parallel postulate. There simply was no safe method by which the security of formal mathematical systems powerful enough to represent the natural numbers could be ensured. And there simply was no such thing as a formal system that could adequately and completely represent the natural numbers. Gödel's mathematical methods, too, found their way into daily mathematics, including the employment of recursive functions, the arithmetization of metamathematics, and the construction of "Gödel formulas" to establish, via self-reference, the incompleteness of formal systems. Never again would syntax be substituted for semantics, proof for truth. But the wider significance of Gödel's achievement, its true meaning, was something else.

One would have thought that after Gödel's incompleteness theorems, which established the essential limitations of formalization, the very enterprise of formalizing mathematical domains would have been reconsidered. Yet nothing of the kind happened. The American logician Emil Post was one of the few to take note of this curious fact. In an essay submitted to a mathematical journal in 1941 (and rejected), Post observed that as a result of Gödel's theorems we know that "mathematical thinking is, and must be, essentially creative." He went on to remark, however, that "it is to the writer's continuing amazement that ten years after Gödel's remarkable achievement current views on the nature of mathematics are thereby affected only to the point of seeing the need of many formal systems, instead of [a single] universal one." One would have expected, he went on to say, that the fascination with formal systems, with proof and syntax, would give way to "a return to meaning and truth." Yet this never happened. The zeitgeist would not be denied. (Even the members of the explicitly antiformalist "intuitionist" school of mathematics championed by L.E.J. Brouwer would in the course of time expend a great deal of energy constructing formal systems for parts of intuitionist mathematics.)

A more general aspect of Gödel's results remains neglected to this day. "Gödel's program," though closely related to Hilbert's, was different in an important way. Whereas Hilbert wished to do for certain parts of mathematics what the positivists wished to do everywhere—replace the intuitive with the formal—Gödel's overarching ambition throughout his career consisted in the attempt to establish, by formal means, the limits of formal methods in capturing intuitive concepts. The goal of his incompleteness theorem was thus to establish, by the most formal of methods—methods that could be programmed into a computer—the limits of formal systems of proof in capturing the intuitive concept of mathematical truth. Church's thesis, in turn—which made precise the concept of a formal system, and thus demonstrated that Gödel's results applied to any formal system whatsoever—itself constituted a chapter in Gödel's program, insofar as the intuitive notion of effective calculability or mechanical solvability, an epistemological concept, concerning what we can come to know using mere calculation, was to be identified with the formal, mathematical concept of a (general) recursive function. As Church himself made clear, his thesis was based on an idea of Gödel's, inspired by a suggestion of Herbrand's, that it might be possible to identify the intuitive concept with the formal one. What was remarkable was that here, unlike the case of Gödel's theorem, there turned out to be no essential limitation on the effort to find a formal characterization of an intuitive concept. In this instance, the "Gödel program" achieved a positive result. Gödel himself found this astonishing. With the concept of [general] recursiveness or Turing computability, he said later, we have "for the first time succeeded in giving an absolute definition of an interesting epistemological notion." It was, he said, "a kind of miracle" that the diagonal procedure (as in his incompleteness theorem) "does not lead outside the defined notion."

What was striking in the case of Gödel's incompleteness theorems, however, was not just that his program was satisfied negatively, but that this result was proved formally. (Church's thesis, though convincing to most, including, eventually, Gödel, cannot be proved formally, since one of the concepts involved, effective calculability, is of course precisely not formal, but rather, intuitive.) That is what made his results so irresistible and so aggravating. This was the secret of Gödel's strategy: where possible, he would establish the limits of the formal from within the formalism itself. He was a mathematician one of whose principal tasks was proving, mathematically, what formal mathematics can and cannot accomplish.

This overarching methodology of Gödel's, however, while practiced in plain sight, has proved invisible. No one noticed that Gödel's later contribution to relativity theory provided yet another example of his program of discovering the limits of formal methods in capturing intuitive concepts, this time, however, with a reversed conclusion. It was not only Gödel the logician who was a spy. Gödel the philosopher proved to be a figure yet more difficult to discern.

5 | It's Hard to Leave Vienna

In the midst of the exultant joy that is pervading our country . . .
you will be very happy if . . . in accord with the true will of the
Führer you may be allowed to support the decision of his now
united people with all your strength.

ERWIN SCHRÖDINGER, AUSTRIA,
1938, "CONFESSION TO THE FÜHRER"

Having announced his incompleteness results at Königsberg and published them shortly thereafter, Gödel rocketed to international fame in mathematical and logical circles. In the months and years that followed, he presented these results, and a slew of others, in a host of mathematical colloquia, including those presided over by his advisor Hans Hahn and by Hahn's friend and colleague Karl Menger. In Menger's colloquium, in 1933, Gödel delivered a paper on the relationship of classical to intuitionist logic. In the audience was the American mathematician Oswald Veblen, newly recruited by Abraham Flexner to join the mathematical faculty of the Institute for Advanced Study in Princeton, New Jersey. Veblen, impressed, assisted Flexner by doing some recruiting of his own.

Abraham Flexner Goes Shopping

In 1929, just moments before the stock market crashed, the owners of the New Jersey department store Bamberger's, Louis Bamberger and his sister Mrs. Felix Fuld, sold the business to R.H. Macy and Co. with the intention of using their profits to found an institution of higher learning. To assist them in this endeavor they enlisted the services of Abraham Flexner, a distinguished educational reformer. Flexner proposed an Institute devoted entirely to the exercise of pure thought of the highest order, where the faculty, the elite of the elite, would be unencumbered with the usual burdens of teaching and administration. (The burden of publication, of course, would remain.) He quickly convinced Bamberger and Fuld that this Institute for Advanced Study should be located in Princeton, New Jersey, where it could take advantage of the resources and the tradition of academic culture provided by Princeton University. Flexner argued as well that the institute's first and fundamental faculty should be in the area of mathematics—including mathematical physics—which combined purity, significance, and universally accepted canons of objectivity. Wasting no time, he began by recruiting, in 1932, Oswald Veblen and Albert Einstein as permanent full professors. Veblen he stole from Princeton University, the first of several thefts that would strain the relationship between the two institutions. A vigorous and enthusiastic shopper with an eye for geniuses, within a year Flexner acquired the mathematicians Hermann Weyl, James Alexander, and John Von Neumann.

Getting Einstein, clearly, was a coup. In California in 1931, trolling the halls at Caltech for his new institute, Flexner learned that Albert Einstein happened to be visiting. The great physicist was intrigued but cautious, having already received offers from prestigious and well-established centers of learning around the world. He advised Flexner to approach him again in Oxford, England, where he would be spending the spring semester of 1932. Flexner did, but Einstein remained ambivalent, suggesting that Flexner approach him again that summer, when Einstein would be at his country home in Caputh, Germany. The third time was the charm. Ambivalence was replaced with enthusiasm: "I am fire and flame for it," announced Einstein.

If Flexner's shopping trip had been productive, so was Veblen's. A guest at Menger's mathematical colloquium in Vienna, 1933, Veblen was so impressed with Gödel's presentation, which confirmed, no doubt, what he had already heard from others, that he invited the young logician to visit the Institute for Advanced Study during its inaugural year, 1933–1934. Gödel would not be a permanent professor, like Veblen himself, but one of twenty-four "workers," as they were then called. (They would soon acquire a more genteel title, "temporary member.") Veblen enticed Gödel by offering him the opportunity to attend a seminar on quantum mechanics to be offered at the Institute by Von Neumann. Gödel responded that he had a "lively interest" in quantum mechanics and would welcome the opportunity. (His interest in physics was indeed lively. He returned from the library in those years with volumes by Schrödinger, Dirac, Planck, Mach, Born, and Lorentz.)

If he came to Princeton, Veblen pointed out, Gödel would have the opportunity to work with the American logician Alonzo Church, who had taken his doctorate under Veblen and who had just developed a new system of formal logic, including what would later be known as the lambda calculus. It would be interesting, Veblen hinted, to see whether Gödel's incompleteness theorem applied with equal force to Church's system of logic. Gödel agreed that it would indeed be very interesting. So, it turned out, did Church, who believed, however, that his formal system was sufficiently different from the one Gödel had focused on in his proof, the *Principia Mathematica* of Russell and Whitehead, that it would be untouched by Gödel's reasoning. (Church, it turned out, was mistaken. Worse still, his own students, Kleene and Rosser, proved his system inconsistent. In time he would become a champion of attempts to prove Gödel's results inescapable by any genuinely formal system.) A third benefit that would attend a visit was suggested by Gödel himself: the opportunity to improve his English. The offer was too good to refuse. So, in the course of time, was a second invitation, and then a third. Yet in spite of the nightmare that was the political scene in Austria, Gödel remained hesitant about a permanent move to the institute and America. Even on the eve of the Nazi occupation (a conquest welcomed in the streets with patriotic fervor), he strengthened his commitment to Vienna.

The Vienna Syndrome

Other Viennese intellectuals suffered from the same syndrome: an attachment to the city of charm and culture so unreasonably strong that even the rumblings of the approaching German war machine could not dislodge them. In its worst hour, the German-speaking world of Austria-Hungary and Germany still offered such intellectual depth and warm collegiality to likeminded thinkers that its luminaries feared, perhaps rightly, that nowhere else would their light ever again burn so bright. Erwin Schrödinger, one of the pioneers of quantum mechanics, certainly thought so. Having abandoned Germany in the tumultuous early 1930s, he became disenchanted with life in Britain and America and decided in 1936 to return to his native Austria, where, and only where, he could flourish. He would later describe this decision as "an unprecedented stupidity."

Having deliberately climbed into a hole, Schrödinger proceeded to dig it deeper. A self-advertised apolitical man, he was especially allergic to Nazis. Newly appointed to the Karl Franzen University of Graz, he couldn't resist the opportunity, in giving a lecture in Vienna on "World Structure in the Large and in the Small," introduced by Hans Thirring, who had taught physics to Gödel, to append at the conclusion a political remark. "When one returns again from the kingdom of the stars," he said, "to our world, one finds there a liking for a concept that wants to place one of the nations that live in this world over or under another one."

The audience cheered wildly; the Nazis took note. When the Anschluss occurred, it became clear that if he were to remain at Graz, Schrödinger would have to make an equally public retraction. This he proceeded to do in a "Confession to the Führer" on March 30, 1938, published in all the Austrian and German papers. "In the midst of the exultant joy that is pervading our country," he wrote, ". . . you will be very happy if . . . in accord with the true will of the Führer you may be allowed to support the decision of his now united people with all your strength." This embarrassing statement he tried to explain afterward to his friend Albert Einstein. He knew, of course, that his return to Austria would be dangerous, but he had not anticipated that "the fortress would be surrendered without striking a blow." "I hope you have not seriously taken amiss," he wrote to Einstein, "my certainly quite cowardly statement . . . I wanted to remain free and could not do so without great duplicity." The duplicity paid off, but its effect was brief. Within months he received a short note from the interior ministry dismissing him from the university, "with no right to any legal recourse to this dismissal." The reason for the action: "political unreliability." An embarrassment had turned into a disaster.

Not all who suffered from the Vienna syndrome were so unsuccessful. The Wittgenstein family was one of the richest and most prominent in Vienna. Assimilated Jews for generations, they like many others were awakened by the Anschluss to the fact that in the eyes of the new regime they were not Austrians or Germans but Jews. Ludwig Wittgenstein, the philosopher, was safely in England, his brother Paul, the pianist, in America, where their sister Margarete would also live out the war. The sisters Hermine and Helene, however, could not be cured of the Vienna syndrome. Nazis or no Nazis, the city was their home, and they were not about to leave it. The problem was that the Nuremberg laws declared Jewish anyone with at least three Jewish grandparents, a criterion the Wittgenstein sisters appeared to satisfy. There was, however, a way out. The Nazi regime had a procedure for reclassification of Jews, a *Befreiung*. The Wittgensteins were of "mixed race," *Mischlinge*, but if they could prove that in fact they had only two Jewish grandparents, then—assuming they had not been so unwise as to actually practice Judaism or to marry a Jew—they could be relabeled "*Mischlinge* of first degree" and permitted to remain, albeit tenuously, in the new Reich.

Successful attempts at *Befreiung* were extraordinarily rare. The year the Wittgensteins made their attempt, 1939, saw 2,100 petitions for reclassification. The Führer approved twelve. It seemed, however, that the Wittgensteins had a chance, since service in the Great War counted toward reclassification, and Ludwig and Paul had accumulated their share of medals in service to the fatherland (Paul losing his right arm in the process, Ludwig working on the *Tractatus* in his spare time in the army). But a case would have to be made to the Führer, who declared himself deluged by requests forwarded by the interior ministry: "I get buckets and buckets of such applications . . . my fellow party members! Obviously you know of more decent Jews than there are Jews in the whole of the German Reich." In the end, however, Ludwig and Paul were able to assist their sisters. Medals did not impress the Reich, but gold did, and the Wittgenstein wealth, much of it invested abroad, came to the attention of the German ministry. With the active assistance (if not the approval) of the brothers Paul and Ludwig, and a Viennese lawyer who specialized in such things, Dr. Arthur Seyss-Inquart, who would later be hanged as a war criminal, it was arranged, after extensive negotiations, to transfer a vast amount of the Wittgenstein fortune into the coffers of the Reichsbank in Berlin. In exchange, the ministry agreed to accept as true the story that the Wittgensteins' paternal grandfather, Hermann Christian, was in fact the illegitimate son of the "princely" house of Waldeck, leaving the Wittgenstein sisters with just two Jewish grandparents and permitting their official reclassification as Mischlinge of first degree. No price, it seems, was too high to pay if it enabled the Wittgenstein sisters to remain in Vienna. Hermine and Helene survived the Vienna syndrome without having to undergo a cure.

The Viennese disease also had a German strain. It unmistakably stamped Wilhelm Furtwängler—cousin of Philip Furtwängler, the mathematician whose lectures had so inspired Gödel—legendary conductor of the Berlin Philharmonic, with whom Paul Wittgenstein had once performed with left hand alone. Forced to dismiss musician after talented musician from his beloved philharmonic to satisfy the demands of racial purity, compelled for the same reason to delete composers from Mendelssohn to Mahler from his concert programs, Furtwängler at no point considered the possibility of abandoning his post or his fatherland. He believed it his duty to preserve the sacred legacy of German music, even—indeed, especially—in a time of political madness. His contempt for Nazis, his unceasing attempts to assist Jews—"Can you name me a Jew on whose behalf Furtwängler has not intervened?" bemoaned Josef Goebbels—put him on a permanent short list for the concentration camps. On the knife edge of execution throughout the war, he was preserved only by the patronage of his greatest fan, Adolf Hitler. The attempt by the Nazis to replace Furtwängler with a brilliant young upstart named Herbert von Karajan—who enthusiastically joined the Nazi party not once but twice—fell on deaf ears; the suggestion was dismissed by Hitler with contempt.

Hated by many Germans for his opposition to the Nazis, Furtwängler was yet more hated by the allies for his refusal to leave Germany, a twin fate he shared with his compatriot, the physicist Werner Heisenberg. Heisenberg was, with Schrödinger, a founding father of the "new" quantum theory: Heisenberg of the abstract, algebraic matrix theory beloved of positivists, Schrödinger of the more metaphysical and intuitive wave mechanics. (It was Gödel's friend Von Neumann who would prove them, in a strong sense, equivalent.) Between the Austrian and the German no love was lost. "I . . . felt discouraged, not to say repelled, by [Heisenberg's] methods of transcendental algebra," wrote Schrödinger, "which appeared very difficult to me, and by the lack of visualizability." Heisenberg, for his part, complained that "the more I reflect on the physical portion of Schrödinger's theory the more disgusting I find it. . . . What Schrödinger writes on the visualizability of his theory . . . I consider trash." Yet they shared more than their creation of quantum mechanics, for Heisenberg too suffered from the Vienna syndrome, albeit, like Furtwängler, from the German strain. He too could not bring himself to leave his fatherland during the war, despite his contempt for the Nazis and all they stood for, despite his longstanding admiration for the "Jewish physics" of his friend and hero Albert Einstein, which rendered him, like Furtwängler, a German outcast in the fatherland. At war's end, Heisenberg, like Furtwängler, was a hunted man in his own country; to the outside world, a traitor.

Red Roses from Dr. Einstein

Einstein himself appeared to be immune to the Vienna syndrome. Appearances, however, can be deceptive. Though his exit from Germany in the early 1930s, before many of his colleagues realized it was time to leave, has been widely noted, what has achieved less attention is Einstein's decision in the first place to take up residence in Berlin after his epoch-making papers in the "miraculous year" of 1905 catapulted him to world fame. (This neglect has begun to be corrected.) Having taken the extraordinary step of abandoning his German homeland in his teens to take up residency and citizenship in the less bellicose and academically rigid domain of Switzerland— where he received his degree in physics, where in seven years as a patent clerk he established himself as "the new Copernicus," and where he became established as a professor in Zurich—why did he decide in 1913 to return not just to Germany, but to Berlin, the very heart of Prussia?

Max Planck and Walther Nernst approached the young Dr. Einstein in Zurich and gave him a day to think over their offer of a professorship in Berlin. He would meet them at the train station, bearing white roses if he declined, red if he accepted. The roses were red. "It is not entirely clear why Einstein accepted the invitation to Berlin," writes his biographer Albrecht Fölsing, nor why he chose to stay there after the Great War for which he held his own country to blame. Fölsing offers some suggestions: Planck was after all the first to recognize, and publicize, the importance of the theory of relativity; the position in Berlin was free of any teaching obligations; Elsa Löwenthal, Einstein's cousin and soon his paramour, would be there. Each persuasive, but none conclusive. He seems to have come under the German strain of the Vienna syndrome. This would help to explain why Einstein remained in Germany during and after World War I, when he was a lone pacifist in a sea of not merely German citizens but Prussian scientists caught up in the hysteria of the war, and why he continued to reside in Germany during the early years of the Nazi madness.

The handwriting, after all, had been on the wall a long time. As early as 1920, an "antirelativity," i.e., "anti-Einstein," club had been formed in Berlin, bearing the name "Study Group of German Natural Philosophers." This group was devoted to denouncing Einstein's "Jewish physics," even offering money to those who would join their cause. On August 24, 1920, the insanity found a home. In Berlin's Philharmonic Hall, where Furtwängler had conducted the Berlin Philharmonic and Paul Wittgenstein had performed concertos for the left hand, the antirelativists staged a meeting to make their voices heard. Einstein himself attended, and amazingly found the time afterwards to publish a response. The debate was a tar baby, and Einstein, together with his unhappy friends, became chagrined that he had allowed himself to be drawn into this fray. Yet it would take well over a decade's worth of such events to induce Einstein to leave Germany. The Vienna syndrome indeed, in its Berlin strain.

The Eleventh Hour Plus One

If Einstein left Berlin not quite at the eleventh hour but perhaps at the tenth, Gödel left Vienna not at the eleventh hour but even later, just before midnight. He would visit the Institute for Advanced Study not once but three times before finally making the decision to leave Vienna. He accepted invitations to be a "worker" at the IAS from 1933 to 1934, 1935 to 1936, and 1936 to 1937. Though the visits were on the whole successful—with the exception of the second, which Gödel terminated abruptly, citing health reasons—the years immediately following his great discoveries took a heavy toll on his mental and physical well-being. The lengthy ocean voyages to the States were a great strain, and the work itself (which he engaged in excessively, quickly producing a string of important mathematical results) was also clearly a burden; but the need to defend his work against critical misunderstandings no doubt had the most devastating effect. Unlike Einstein, who would have toasted Franklin Roosevelt's statement that "I am an old campaigner and love a good fight," Gödel throughout his life was deeply averse to dispute and controversy. This had an unfortunate effect on his published papers, which were carefully designed to put forward not all that he knew or believed but only what he could establish beyond all reasonable doubt, what even his opponents would be forced to accept. The manuscripts of his publications, sadly, contain vast amounts of valuable material crossed out, no doubt because they failed to meet his excessive standards of acceptability without controversy. In the end, ironically, this strategy of safety failed completely. All along, Gödel was understood to be a man apart from his times, whose beliefs on a host of topics—from truth and proof, to language-centered philosophy, to God and spirits—were wildly out of step with those of his contemporaries.

With fame came not only friends but enemies, and this Gödel was unprepared for. It was during these years that he suffered his first bouts of depression and began to show signs of the eating disorder that would eventually kill him. Three visits to sanatoria are documented: 1934 at Sanatorium Westend in Purkersdorf, near Vienna; 1935 at Breitenstein am Semmering, and 1936 at a sanitorium at Rekawinkel, near Vienna. These inner struggles were complemented by the outward turmoil of his beloved Vienna, which was coming down around him in ruins. It was in these years that Germany invaded Czechoslovakia and that Austria succumbed to the Anschluss, in this period that Schlick, of the Vienna Circle, was murdered and Herbrand fell to his death while mountain climbing. In addition, Gödel's teacher Hahn had also died, and his friends Menger and Carnap had left for America. He also had more immediate concerns. His mother had been forced by financial circumstances to return to Brno, where her vociferous objections to the Nazis put her at considerable risk.

Gödel himself was eerily silent about the political events surrounding him, both in person and in correspondence, which resulted in a considerable cooling of his friendship with Menger. If this silence sprang from Gödel's excessive caution, then once again it was ineffective. The Nazi authorities noted his association with "the Jewish professor Hahn," adding that "it redounded to his discredit" that he "traveled in liberal-Jewish circles." The signs were all around him. His cleaning lady presented him with a bill at the bottom of which were appended the words, "Heil Hitler!" Yet in spite of everything, he made no decision to leave. On the contrary, as late as 1939 he and Adele, whom he had married in 1938 in a civil ceremony—surprising his brother, who had only just been introduced to her—moved from a rented apartment to one they had purchased in the city of Vienna. The Vienna syndrome, in spades. More still needed to happen before Gödel could be persuaded to flee.

More did. One day late in 1939, Gödel was accosted by a gang of youths who took him for a Jew—or at least someone who associated with Jews—and roughed him up, knocking off his glasses. Adele, fortunately, was able to beat them off with her umbrella. This incident, however, seems to have done the trick, especially when combined with the order he had recently received to take a physical examination to determine whether he was fit for military service with the German army, an examination which, amazingly, the frail, thirty-two-year-old logician, who had been in and out of sanatoria, passed. With the considerable assistance of the new director of the IAS, Frank Aydelotte, who wrote to the authorities that Gödel should be given special treatment as an Aryan who was a world-famous mathematician, Gödel succeeded, in December 1939, in securing a visa to travel to America. The great logician escaped from Austria just as the door was closing behind him. "I am told in all steamship bureaus," he wrote to Aydelotte, "that the danger for German citizens to be arrested by the English is very great on the Atlantic." He decided, therefore, to travel with Adele via the trans-Siberian railway to Japan, from there to voyage across the Pacific to San Francisco, and thence by rail to Princeton. A grueling journey about which Gödel, typically, made no comment. Adele, however, remarked afterward that they traveled frequently at night, in constant fear of being detained and returned to Austria.

The journey itself, however, was without incident, and Gödel arrived at last in San Francisco harbor on March 4, 1940. Soon he would be at the IAS with another survivor, Albert Einstein, whose friendship would be a watershed in two lives that had already marked some of the greatest intellectual achievements of the twentieth or any other century. Gödel and Einstein would discover that what they had left behind in Vienna and Berlin they would never find again. Each in his own way would become increasingly isolated and lonely, a creature of another time and another culture, their native language a constant reminder of their origins in the land that had become their adopted country's mortal enemy. A historic empire had dwindled to the company of two, what Kurt Vonnegut would describe in his novel *Mother Night* as "das Reich der Zwei." But what a two! While lesser souls might look only to the glories of their past, Gödel and Einstein in their remarkable friendship would explore a new world of ideas. More and more, their thoughts would turn to a topic at the very center of Einstein's relativity, at the core as well of Gödel's preoccupation with Kant and Leibniz, the German idealists. The question was time. The answer would be yet another surprise from Dr. Gödel.

Participants at Einstein's 70th birthday celebration. Left to right: Eugene Wigner, Hermann Weyl, Gödel, I.I. Rabi, Einstein, Rudolf Ladenburg, J. Robert Oppenheimer.
Photo by Howard Schrader

Kurt Gödel with his brother, Rudolf, and his mother, Marianne.
Courtesy of the Gödel Archives, Firestone Library, Princeton University, and the Institute for Advanced Study

Kurt and Adele Gödel dining outdoors.

Courtesy of the Gödel Archives, Firestone Library, Princeton University, and the Institute for Advanced Study

Kurt and Adele Gödel dining indoors.

Courtesy of the Gödel Archives, Firestone Library, Princeton University, and the Institute for Advanced Study

Gödel, dressed for driving.
Courtesy of the Gödel Archives, Firestone Library, Princeton University, and the Institute for Advanced Study

Gödel, dressed for hiking.
Courtesy of the Gödel Archives, Firestone Library, Princeton University, and the Institute for Advanced Study

Young Adele Porkert at the piano.

6 | Amid the Demigods

When pygmies cast such long shadows, it must be very late in the day.
GIAN-CARLO ROTA

Princeton is not Vienna. Having fled Nazi-occupied Austria, Gödel and Adele found themselves in an Ivy League college town, small, provincial and inbred, dominated by the imposing presence of Princeton University, itself outclassed by the still more prestigious Institute for Advanced Study. Princeton's students may have touted themselves as the crème de la crème; the institute could boast that mere students, of whatever caliber, were not welcome. In these streets it was hard to avoid rubbing shoulders with the intellectual elite, and with those who thought of themselves as such. Bertrand Russell was unimpressed. He found Princeton "full of new Gothic, and . . . as like Oxford as monkeys can make it." Einstein was more delicate: "Princeton is a wonderful piece of earth and at the same time an exceedingly amusing ceremonial backwater of tiny spindle-shanked demigods." The pygmies on stilts would have to make way for the entrance of two giants.

Gödel, however, was content with his new home. Unlike Adele, who struggled with English, he had long been fluent, and he found the people and culture in Princeton "ten times more congenial" than those in Vienna. He admired also the "prompt functioning of government officials in America," which he said made life "10 x 10 x . . ." better than in the old country. The very fact that intellectual life was more narrowly focused, centered at the institute on mathematics and mathematically oriented sciences, could not fail to please him. But in such a rarefied atmosphere, it is no surprise that Adele, lacking academic pretensions, found it hard to breathe. In the café, opera, club, and cabaret scene of prewar Vienna, the easy charm of social life had blended smoothly with the life of the mind, professors mingled with dancers, and composers dined with philosophers. In Princeton, such mixing would be hard to imagine. (There were, in any case, few cabarets or opera houses to be found.)

Without children, Adele sought refuge in a menagerie of pets that would eventually encompass a pair of love birds, a dog named Penny and a cat, though not one of the tailless Manx variety of which she had become inordinately fond. (Only the desperate pleas of her friend and neighbor, Bobbie Brown, dissuaded her from removing the tail of the cat she did purchase.) In time, the Gödels would also sponsor a foster child abroad. Their first home in Princeton was an apartment at 245 Nassau Street. Later, they moved into an apartment at 108 Stockton Street, where their neighbor, George Brown, already knew Gödel. During one of Gödel's early visits to the institute, Brown, a graduate student in mathematical statistics who had studied logic at Harvard with W.V.O. Quine, had been given the task of taking notes on Gödel's lectures to prepare them for publication. He and his wife Bobbie quickly befriended the new arrivals. But during their visits with the reclusive Gödels, they were put off by their hosts' decision to remove the screens from the windows. Gödel claimed that this allowed him to breathe properly, but it also allowed easy access to dust and insects, which considerably dampened Bobbie Brown's enthusiasm for coming over.

This residence, too, was temporary. The Gödels moved again, this time to an apartment at 120 Alexander Street, near the train station. Gödel was fond of his new abode, remarking that it was located "directly opposite the most elegant hotel in town" (the Princeton Inn). They occupied the entire upper floor of the building, with windows on all sides, which Gödel found helpful in surviving the hot Princeton summers. Adele, however, was unhappy with the poor condition of the premises, which she considered unhygienic, and she found the neighborhood dreary.

Her disappointment turned to enthusiasm when she discovered a house for sale at 129 (later 145) Linden Lane, at the outskirts of town. Built just a few years earlier, it was a small one-story structure of "sturdy cinderblock" with an automatic oil furnace and built-in air conditioning, and included a wood-burning fireplace. Adele would not rest until she had persuaded her husband to purchase it. The house was beyond their means, but Gödel was able to secure a mortgage for three-fourths of the purchase price, while the director of the institute, J. Robert Oppenheimer, arranged for a salary advance to cover the rest. It was, to Gödel's mind, a "somewhat shaky" arrangement, but he went along with it, and in August 1949 they moved in to stay. Adele arranged the furnishings, which were for the most part modest, but she did indulge her weakness for oriental rugs and chandeliers. Over the years, she would oversee the construction of a sun porch, refinish a room to serve as Gödel's study, and plant a flower garden. Evergreens would also be planted in front of their home, as a barrier to shield the Gödels from passersby. Adele engaged in traditional domestic activities. She was, friends noted, a good cook, though with a penchant for heavy German fare. In the summertime, she was relieved to escape the confines of Princeton by accompanying Gödel to resorts in Maine.

Still, it was a sad life, centered on the principal occupation of tending to her fragile husband, a task Adele shared with a succession of Princeton luminaries. In the early years it was Oswald Veblen, the one who had first moved to acquire Gödel for the institute, who took it on himself to look after him. The job then passed to Einstein, and after Einstein's death to the economist Oskar Morgenstern. Shepherding the logician became increasingly necessary, for Gödel's eccentricities, already evident in Vienna, had blossomed to such an extent that as early as 1941, Frank Aydelotte, director of the institute before Oppenheimer, took the extraordinary step of inquiring of Gödel's doctor whether "there is any danger of his malady taking a violent form

which might involve his doing injury either to himself or to others
. . . " He was assured that violence would not be a problem.

These eccentricities included Gödel's conviction that "bad air" was
emanating from his refrigerator as well as from the heating system.
Fear of air was accompanied by an aversion to cold. Even in midsum-
mer, his gaunt figure could be seen enshrouded in a winter coat, along
with hat and gloves. His nascent hypochondria developed into a full-
blown syndrome in which imaginary maladies were given the same
status as real ones. Worse, fear of disease was accompanied by fear
and mistrust of doctors. There was also Gödel's increasingly reclusive
behavior. He endeavored to abstain from all "unnecessary" social and
intellectual interaction and was at pains to avoid, at all costs, actual
physical contact with other human beings. The telephone became his
preferred method of communication. (His penchant for waking up
long-suffering friends at all hours of the night to engage in endless telephone calls puts one in mind of another reclusive genius, the Canadian
pianist Glenn Gould, who could also be seen in the summer heat bun-
dled up in coat, hat and gloves.) Fear of disease finally grew into a generalized fear of others, a paranoia that may or may not have evolved
into psychosis. The significance of his mental instability had been presaged by his great mathematics professor at the University of Vienna,
Philip Furtwängler, who, on learning of Gödel's incompleteness theorems, asked, "Is his [mental] illness a consequence of proving the unprovability [of the consistency of formal arithmetic], or is his illness
necessary for such an occupation?" Gödel's eccentricities would eventually contribute to his death, cited by his doctor as due to "inanition"
brought on by personality disorder.

Einstein's eccentricities also flourished in his new homeland, but
they were of an altogether milder order and affected his family and
friends far more than himself. The great scientist's colleagues were embarrassed by his unprofessorial penchant for strolling down Nassau
Street, Princeton's leafy main thoroughfare, while licking an ice cream
cone, and female friends were taken aback by his sexual frankness and
disdain for traditional mores. More generally, Einstein's attitude to-

ward women was nothing to boast of. Peter Bucky, who had plenty of opportunities to view the great physicist up close while serving as chauffeur and companion for his father's friend, has written that "in today's terms, Einstein would have been considered a classic male chauvinist. He once wrote a letter to a friend, a Dr. Muesham in Haifa, that his definition of a good wife was someone who stood somewhere between a pig and a chronic cleaner." In Princeton, a bastion of bourgeois intellectual respectability, Einstein raised his bohemian lifestyle to an art. His unkempt hair became longer with each succeeding year. He wore shoes without socks and a leather jacket, not from a sense of fashion but from the belief that what doesn't wear out won't need to be replaced. His son Hans, who came to America already estranged from his famous father, was especially embarrassed by Einstein's manner of dress and became, in protest, a dandy. The complete neglect of all surface appearances may be the signature of a deep thinker, but it does little to smooth over the already fractious relations among human beings. The one companion that suffered no harm was the omnipresent pipe. His friend Gariella Oppenheim-Errara recalled an occasion when his sailboat, *Tannef*, capsized, leaving the great physicist paddling in the water, puffing contentedly on his faithful pipe. There is no doubt that Einstein's pipe was his closest associate, while others—including wife and family—were never permitted the illusion that they would ever be at the center of his life.

Einstein the bohemian was also a Jew in a town whose university still enforced Jewish quotas for its students, something Einstein had personally protested during the negotiations for his appointment. Caught in this net was Richard Feynman, one of the great physicists of the modern era, who was forced to attend MIT due to Columbia University's quota system. We can infer that his acceptance at Princeton University for graduate studies was also subject to ethnic strictures: Philip Morse, Feynman's professor at MIT, felt compelled to note in his recommendation that Feynman's "physiognomy and manner . . . show no trace of this characteristic [i.e., being Jewish] and I do not believe the matter will be any great handicap." (Feynman would go on to become the third winner of the Einstein Prize, after Gödel and Schwinger.) In this environment, as an eccentric Jewish professor with a bohemian lifestyle and a heavy German accent, who moreover had devoted much of his life to a doomed search for an elusive theory that would unite relativity with quantum mechanics, Einstein was unquestionably an outsider. Clearly, he would appreciate the appearance in Princeton of another outsider whose eccentricities, if not genius, surpassed his own.

It was in 1933 that Gödel first met Einstein, during a visit to the institute, when they were introduced by another émigré, Paul Oppenheim. They became friends in 1942, after Gödel, too, had joined the institute, and remained close until Einstein's death in 1955. When Gödel was ill in February 1951, he wrote to his mother, Marianne, that "during my sickness Einstein was of course extraordinarily nice to me and visited me many times both in the hospital and at home." The exact circumstances surrounding the transition from acquaintance to friendship remain unknown, and not by accident. The two men cherished the privacy of their relationship. After his good friend died, Gödel wrote to his mother that the fact "that people never mention me in connection with Einstein is very satisfactory to me (and would be to him, too, since he was of the opinion that even a famous man is entitled to a private life). After his death I have already been invited twice to say something about him, but naturally I declined."

He was happy, however, to confide some details of their friendship to satisfy his mother's curiosity. Gödel, it emerged, would meet Einstein at his home each day between ten and eleven in the morning, and the two would walk to the institute, a journey that generally took half an hour. At one or two in the afternoon, they would return home, discussing politics, philosophy and physics. This schedule gives new meaning to Einstein's comment that he went to his office "just to have the privilege of walking home with Kurt Gödel." These walks, in fact, consumed some thirty percent of his workday. They could also be dangerous. "I know of one occasion," wrote Einstein's secretary, Helen Dukas, in 1946, "when a car hit a tree after its driver suddenly recog-

nized the face of the beautiful old man walking along the street."
Gödel's more severe countenance, by contrast, was no threat to traffic.
"I have so far," he wrote to his mother, "not found my 'fame' burden-
some in any way. That begins only when one becomes so famous that
one is known to every child in the street, as is the case of Einstein."

Gödel, by nature a pessimist about human affairs—though an optimist about the power of reason—was balanced by the more opti-
mistic Einstein. Yet Einstein too, as Gödel wrote to his mother, "was
in many respects a pessimist. In particular, he didn't have a very good
opinion of humanity in general. Among other things, he based this on
the fact that those who wished to do some good, like Christ, Moses,
Mohammed, etc., either died a violent death or had to use violence
against his followers."

Both men were skeptical of Bohr and Heisenberg's Copenhagen interpretation of quantum mechanics, and Gödel was skeptical as well of
Einstein's efforts to unify relativity with quantum physics. Each lived
in a modest home, while colleagues like Von Neumann inhabited man-
sions. Their households mingled; they exchanged housewarming gifts.
Their lives became interconnected. It was a familiar sight in Princeton
to see the two friends walking home from the institute, arguing in their
mother tongue about politics and general relativity. Einstein, like many
intellectuals, favored Adlai Stevenson for president. "Gödel," however,
Einstein remarked, "has really gone completely crazy. He voted for
Eisenhower." Whoever came to know the one, ipso facto became acquainted with the other.

Submarines Again

Their most famous discoveries behind them, Einstein and Gödel led
increasingly quiet lives in the backwaters of their respective fields. In
domains they had once ruled as titans they were now but part of the
furniture, albeit, as Einstein cracked, "museum pieces." "In Princeton," he told friends, "I am known as the village idiot." As the war

years dragged on and the light of science prepared to cast the shadow of the atomic bomb, the two brightest minds in the scientific firmament drifted ever farther from the center. A few doors away from Einstein's office in Fine Hall, Neils Bohr and John Wheeler were working out the details of nuclear fission for employment in a nuclear device. Einstein, meanwhile, alone and all but inaccessible, pursued his dream not of splitting the atom but of unifying physics. Visiting the institute a few years earlier, the father of the atomic bomb, J. Robert Oppenheimer, commented that "Einstein is completely cuckoo." The gods, as is well known, love irony. After the war, Oppenheimer found himself director of the institute and thus, nominally at least, Einstein's boss.

In 1935, the year Oppenheimer paid his visit to the institute, the father of relativity announced that in his view, the idea of constructing a bomb by splitting the atom was as promising as "firing at birds in the dark, in a neighborhood that has few birds." Yet Einstein would eventually coauthor a letter to Roosevelt suggesting that the prospects for an atomic bomb be explored, becoming thus, in effect, the father of the father of the bomb. Einstein's paternity, however, was at most symbolic. The causal efficacy of his letter to Roosevelt appears to have been minimal. Nevertheless, the sometime pacifist put his fingerprints on the most lethal weapon ever devised by man, and he would spend the rest of his life preaching against the deployment of the weapon he had urged Roosevelt to develop.

While the cream of the physicists and mathematicians, many of them close associates of Einstein and Gödel, including Von Neumann—whose brilliance stood out even among the cluster of luminaries at Los Alamos—had assembled for the bomb project in Oppenheimer's back yard in New Mexico, Gödel and Einstein remained behind in Princeton, lost in clouds of abstruse mathematics and philosophy. Gödel had embarked on an intellectual task that would prove as elusive as Einstein's search for a unified field theory. He took up the quest, inaugurated by Georg Cantor, of determining the cardinality of the continuum—in plain English, counting the number of points on a line. (It was this that had preoccupied Gödel on his nocturnal sojourns during his summer break at Blue Hill in Maine, not, as nervous locals thought, assisting German submarines prowling off the rugged coast.) Everyone knew, of course, that the number of points on the real number line is infinite, but after Cantor's epoch-making discovery that infinity itself came in different sizes, the hunt was on to discover the exact size of this infinity.

Cantor had begun the quest in 1878 with his "continuum hypothesis" in which he speculated that the number of points on a line, $2^{\aleph_0}$, is the very next infinite number, $\aleph_1$, after the smallest infinite number, $\aleph_0$, the cardinality of the set of all natural numbers. He thus hypothesized that $2^{\aleph_0} = \aleph_1$. But he died defeated, having failed to prove his conjecture. Plagued by bouts of depression, he was several times confined to a sanatorium, a sad precedent repeated by Gödel. Gödel, for his part, after much effort, was finally able to prove in the late 1930s and early 1940s that the continuum hypothesis was consistent with Zermelo-Fraenkel set theory—that is, that it could not be disproved. Later, in 1963, a brash young mathematician, Paul Cohen, of Stanford University, who had worked on the problem for two years at the institute, looking down his nose at the petty concerns of mere logicians, succeeded in proving that it could not be proved from these same axioms. Cohen's result is known as the independence of the continuum hypothesis, Gödel's as the consistency of the continuum hypothesis.

Einstein, embracing his doomed search for a unified field theory and deemed a security risk for his sympathies for and contacts in the world of "socialism," discovered that he was not to be trusted with the defense of his adopted land. His opportunity to serve (a lesser one) came not from the desert at Los Alamos but from the sea. The U.S. Department of the Navy engaged him not to design new gyroscopes, as he had once done for his fatherland, but to calculate the explosive potentialities of torpedoes. Calculate he did, but though expensive experiments would later confirm that Einstein's results had indeed been accurate, there is no evidence that his contributions were put into practice during the war.

If Einstein's mathematics did little to advance the war effort, he had greater success in contributing financially. Not, to be sure, by emptying his pockets (which were already empty, as his wife made certain, distrusting his financial competence). Instead he was asked to donate the original manuscript of his 1905 treatise on special relativity for an auction to produce funds to support the war. But Einstein, untouched by manuscript fetishism, had discarded the original. Unfazed, the authorities asked him simply to rewrite it. The amazed professor was happy to oblige, and in due course wrote again his famous paper, this time dictating it to his faithful assistant, Helen Dukas. "Did I really say that?" he interjected from time to time to the unflappable Ms. Dukas. Reality, here, clearly, has competed with fiction. It was Jorge Luis Borges who wrote "Pierre Menard, Author of the Quixote," in which Menard "did not want to compose another Quixote—which is easy—but the Quixote itself." Did he succeed? "Cervantes' text and Menard's," Borges notes dryly, "are verbally identical, but the second is almost infinitely richer . . . more ambiguous . . . " Whether or not Einstein's second writing of his 1905 paper is more ambiguous than the original, it is certainly richer: an insurance company bought the new manuscript for $6.5 million.

It is a sign of both her boredom and her skill as a seamstress that throughout the war years Adele is said to have contributed to the Austrian relief services a dress a day to be given to a young child, in recognition of which she received a bust of her late father from the Viennese authorities after the war. Gödel is not known to have made any direct contribution to the Allies' war effort. He told his friend Atle Selberg, however, that he volunteered after the war ended to serve as a civil defense aircraft spotter. He also proceeded to formalize his commitment to the United States, becoming a citizen of his adopted country in 1947. As witnesses for the ceremony he brought along Morgenstern and Einstein. He had already alarmed the former by confiding to him, in consternation, that he had discovered an "inconsistency" in the Constitution. Apprised by Morgenstern of the danger ahead, Einstein took it upon himself to distract his friend on the way to the swearing in, entertaining him with worn-out jokes and twice-told anecdotes. Einstein might have been even more concerned if he had known that for years the FBI had been intercepting and reading parts of Gödel's correspondence with his mother, who was living in Vienna.

The strategy proved unsuccessful. When judge Philip Forman, who only a few years earlier had ushered Einstein himself into the land of liberty, asked Gödel casually, "Do you think a dictatorship like that in Germany could ever arise in the United States?" he received a spirited reply in the affirmative. Gödel launched into an account of how the United States Constitution formally permitted just such a regime to arise. Shrewdly, however, the judge cut off the great logician before he could hit full stride, and the ceremony came to a peaceful conclusion, leaving Gödel's new homeland to fend for itself against the opening he had discerned in its founding principles. Years later, asked for a legal analogy for his incompleteness theorem, he would comment that a country that depended entirely upon the formal letter of its laws might well find itself defenseless against a crisis that had not, and could not, have been foreseen in its legal code. The analogue of his incompleteness theorem, applied to the law, would guarantee that for any legal code, even if intended to be fully explicit and complete, there would always be judgments "undecided" by the letter of the law.

If Gödel made no explicit contribution to the war, still his mathematical work, in particular his foundational papers on recursive functions—which constitute the soul, if one can put it thus, of the computing machine—would contribute in a profound way to the project of building bigger and bigger thermonuclear bombs, and more generally to the still ongoing program of computer-based military technology. Similarly, while Einstein took no part in the deliberations of Oppenheimer's bomb makers at Los Alamos, his foundational work in relativity formed part of the theoretical background of the very practical results reached in the heart of the New Mexican desert. Gödel and the computer, Einstein and the bomb. Neither man contributed to the technology (or its ethos), but each one's research was essential background for those who did.

Not only did Einstein and Gödel refrain from contributing to the ethos of these technologies, they were completely against it. Their "unfashionable pursuits" gained them fame but not friends. Einstein, pursuing the elusive unified field theory, was a lone figure unwilling to forgo determinism in physical theory or realism in the quantum world. Gödel was a rare spirit keeping the faith that in spite of his incompleteness theorem and the independence result achieved by Paul Cohen, the mathematical universe of sets and numbers was a fully determined, complete reality. Where Cohen led the way for the majority with his belief that there was no more objective truth value attached to Cantor's continuum hypothesis than to Euclid's parallel postulate, only a choice of which convention to follow, Gödel never ceased believing that the comparison with Euclid was misguided. The true axioms that would settle the continuum hypothesis, he felt, were out there to be found. As he wrote to Alonzo Church, who shared Cohen's beliefs, "You know that I disagree about the philosophical consequences of Cohen's result. In particular, I don't think realists need expect any permanent ramifications as long as they are guided, in the choice of the axioms, by mathematical intuition and by other criteria of rationality."

"Einstein and Me": Scientists as Philosophers

The increasingly philosophical turn taken by the two thinkers was becoming more and more unfashionable. In 1935, Einstein, with the assistance of Boris Podolsky and Nathan Rosen, published a kind of philosophical manifesto (which contained what is known popularly as the EPR paradox), "Can Quantum-Mechanical Description of Physical Reality Be Considered Complete?" The authors suggested that the Copenhagen interpretation of quantum mechanics, pioneered by Bohr and Heisenberg, led to paradoxical results, including instantaneous, noncausal action between spatially separated events (which they dubbed "spooky action at a distance"). Not only did this little essay arouse the ire of the quantum-mechanical establishment, including Bohr and Wolfgang Pauli, but on its own terms the document seemed as much philosophy as physics. Einstein brought to bear what seemed to be a priori philosophical considerations concerning the "completeness" of a physical theory—which, the reader was informed, requires the correspondence of each significant element of the theory with an "element of reality"—as well as what constitutes physical reality, "no reasonable definition of which," Einstein insisted, permits what is real in one system to depend on the measurement of another system.

Einstein's philosophical perspective was a form of realism. Gödel too was committed to realism, in the physical as well as the mathematical realm. He believed that mathematical objects and properties exist objectively and independently of knowledge of them by the human mind. He was aware of the parallel between his and Einstein's philosophies. "The heuristics of Einstein and Bohr," he told Hao Wang, "are stated in their correspondence. Cantor might also be classified with Einstein and me. Heisenberg and Bohr are on the other side." His own philosophical manifesto appeared in 1947, disguised as a popular survey of the status of the continuum hypothesis, entitled "What is Cantor's Continuum Problem?" He had been invited in 1946 by Lester Ford, editor of the *American Mathematical Monthly*, to make a contribution to a "What is . . . ?" series whose purpose was to provide an introduction to "a small aspect of higher mathematics" in "as simple, elementary and popular a way" as possible. Gödel, however, took this as an opportunity to formulate a manifesto declaring and defending his mathematical Platonism. As an epistemological adjunct, he introduced the concept of mathematical intuition, for which formal proof was no substitute. "I don't see any reason," he wrote, "why we should have less confidence in this kind of perception, i.e., in mathematical intuition, than in sense perception, which induces us to build up physical theories." According to Gödel, since the continuum is a real object, it was only a matter of time before new axioms would be discovered that would settle the continuum hypothesis, axioms that would "force themselves upon us as being true."

The manifesto, though shocking, was no surprise. The stage had been set by a previous essay, his contribution to the volume on Bertrand Russell for the "Library of Living Philosophers," edited by P.A. Schilpp. This series, inaugurated in 1938, was devoted to great living philosophers at the twilight of their careers. The person to whom a volume was dedicated would contribute an intellectual autobiography, followed by critical assessments of his philosophy provided by leading figures in the field, to be succeeded by the author's responses to those essays. In the volume on Russell, Gödel cited with approval Russell's remark (written before his encounter with Wittgenstein) that "logic is concerned with the real world just as truly as zoology, though with its more abstract and general features." This expression of logical realism Russell himself later rejected under Wittgenstein's influence, a rejection for which Gödel took him to task. Much of Gödel's contribution to the Schilpp volume consisted of a sustained critique of Russell's mathematical philosophy, and he looked forward to Russell's response.

He had reason to look forward to it. In his letter of November 1942 inviting Gödel to contribute to the volume, Schilpp had said that "in talking the matter over last night with Lord Russell in person, I learned that he too would not only very greatly appreciate your participation in this project, but that he considers you the scholar par excellence in this field." In the event, however, Gödel was late in submitting his essay, and Russell claimed that having already responded to the other contributors, he "lacked the leisure" to compose a proper reply to Gödel. Deeply disappointed, Gödel attempted unsuccessfully to change Russell's mind. The only response that appeared in the Schilpp volume was Russell's comment that since it was "eighteen years since [I had] last worked on mathematical logic" it would have taken him "a long time to form a critical estimate of Dr. Gödel's opinions."

Behind this remark lurked an ambivalence on Russell's part toward his great colleague. The incompleteness theorems were not calculated to arouse warm feelings in Russell's breast. As we saw earlier, it is likely that Russell did not fully grasp them. He also claimed that, in any case, he had never subscribed to the dream of the Hilbert school that a consistency proof could be found for his logical system. This would take some of the sting out of Gödel's discoveries, but it would apply only to the second incompleteness theorem. The first theorem, which neither Russell nor anyone else had expected, demonstrated the essential incompleteness of any formal logical system in capturing the truths of mathematics. Its effect, therefore, was to deal a mortal blow to the formalist project of which Russell's masterpiece, *Principia Mathematica*, was the crowning achievement. Gödel had added insult to injury by deciding, in the proof of his theorem, to focus exclusively on *Principia Mathematica* as an example of a formal system that is essentially incomplete. Not only had Russell been surpassed by Gödel as the preeminent logician of their time, his magnum opus had been shown by Gödel to be in an important respect a failure.

If Russell had been eclipsed in logic by Gödel, he was overshadowed in philosophy by his former student Ludwig Wittgenstein. "It is not an altogether pleasant experience," wrote Russell late in life, "to find oneself regarded as antiquated after having been, for a time, in the fashion. It is difficult to accept this experience gracefully." He attributed this change of fashion explicitly to the rise of Wittgenstein, "by whom I was superseded in the opinion of many British philosophers." Not only the content but the very form of Russell's writings was rendered obsolete by Wittgenstein's works, which called to mind the Zeus-like pronouncements of pre-Socratic philosophers such as Parmenides and Heraclitus more than the didactic, quasi-scientific prose that Russell had made his trademark. And Wittgenstein was not the end of it. The final grain of salt in Russell's wounds came from Einstein, whose revolution in physics eclipsed in the popular imagination not only Russell's contributions to logic and philosophy but Gödel's and Wittgenstein's as well. If he had it to do over again, said Russell wistfully at the end of his life, he would have become a physicist.

Triply eclipsed, Russell came to the Institute for Advanced Study in the spring of 1943, while Gödel was composing his critique of Russell's philosophy of mathematics, with a large chip on his shoulder. He noticed Gödel in the audience when he lectured but was so out of touch with recent developments in logic and mathematics that he failed to recognize Von Neumann. When he met for discussions at Einstein's home with Gödel, Einstein and Pauli, it was predictable that Russell would be unsatisfied, as revealed by the barbed comments he committed to his *Autobiography* about "the German bias for metaphysics" of his companions, "all three of them Jews." That his application, several years earlier, for membership at the Institute for Advanced Study, though supported by Einstein, had been declined was the final insult. On the other side, Russell's dismissal of Gödel as an "unadulterated Platonist" did not earn him a warm spot in Gödel's heart.

That the only professional philosopher in the quartet of Russell, Einstein, Gödel and Pauli was disappointed with the others' attachment to the grand old style of "German metaphysics" is a clear sign of the low estate to which philosophy had been reduced. As the last of the great philosopher-scientists, Einstein and Gödel were anomalies. But if the physicist and the logician had both become increasingly philosophical since joining the institute, there remained a difference. For the most part, Einstein's philosophy was immanent in his physics, just as the God of his beloved Spinoza, the great pantheist, was contained in the world. Much of Gödel's philosophy, too, was contained in his mathematics, but he took pains to make some of it explicit—separate, though closely related to his formal discoveries—just as the God of his philosophical hero, Leibniz, was a being apart.

Gödel's desire was to become a great philosopher in the tradition of Plato, Leibniz, and Kant, but he discovered that he had set this goal too late in his life, having devoted his best years to logic, mathematics and physics. The scope of his ambitions—and the degree to which he was at cross purposes with the zeitgeist—can be seen in the fourteen philosophical theses he committed to his notebooks in the 1960s under the title "My Philosophical Viewpoint." "Concepts," he states, "have an objective existence." Along the same lines he writes that "materialism is false." Neither thesis is surprising or unexpected. But he goes further: "The world is rational." This puts one in mind of philosophical theism, according to which the order of the world reflects the order of the supreme mind governing it. Plato, a philosopher Gödel greatly admired, held similarly that all order is a reflection of rationality. Concerning religion Gödel asserts, "Religions are, for the most part, bad, but religion [i.e., belief in God?] is not."

He goes on: "There are other worlds and rational beings of a different and higher kind.... The[se] higher beings are connected to the others by analogy, not by composition." And so on. Gödel was striving for a new picture, a worldview that would put the world into a better perspective than it is at present. From his discussions late in life with Hao Wang, it emerges that he believed that the proper philosophy should capture axiomatically—though not purely formally—the fundamental concepts that underlie reality, which he took to include "reason, cause, substance, accidens [a traditional Latin term], necessity, value, God, cognition, force, time, form, content, matter, life, truth, idea, reality, possibility." Gödel the philosopher, according to Wang, said he wished to "do for metaphysics as much as Newton did for physics."

Einstein, of course, had already done for physics what Newton did for physics. But if his philosophical aspirations did not match Gödel's, the direction of his thought often did. Like Gödel, he was opposed to positivism, but had, when necessary, exploited its weapons to further his own scientific and philosophical objectives. The protopositivism of Einstein's special theory of relativity, which identified time with what could be measured by synchronized clocks, was not a philosophical calling card—though the Vienna Circle seized on it as such—but simply the right tool for the job. Minkowski's generalization of special relativity into the geometrical theory of four-dimensional space-time was a different tool for a different job, which in turn opened the door for Einstein's broader geometrization of space-time in his general theory of relativity, a development that is cold comfort to positivists hoping to adopt Einstein as a true believer. In like manner, with his incompleteness theorem, Gödel exploited the favorite tool of the mathematical positivist, formal systems of proof, to construct a proof that formal systems for number theory will always be incomplete. In essence, the theorem was a mechanical, algorithmic demonstration of the limits of mechanical, algorithmic methods, and as it turned out, of the inescapable limitations of the computer. It was at once a jewel in the crown of formalism and a warning to those who would embrace it.

At the heart of Gödel and Einstein's opposition to positivism was their unfashionable realism, their reluctance to make ontology, the theory of what is, subservient to epistemology, the theory of what can be known. At bottom, the positivist mentality consists in deriving ontology from epistemology. This was the source of Ernst Mach's positivistic objection to atomic theory, since individual atoms will never be directly encountered by humans. But the springs of Mach's philosophy ran deeper. His rejection of a reality "beyond" what appears to human sensibility was a simplified version of Kant's philosophy. The "Copernican Revolution" in epistemology inaugurated by Kant consisted in the radical doctrine that the known must conform to the knower. The hard-nosed, ultraempiricist Mach had derived his positivism from Kant, who was not a realist but an idealist, albeit of the deep, German, transcendental variety, not (as Kant saw it) of the shallow British strain of George Berkeley.

Still, idealism is idealism, whether British or German. Although Kant, unlike Mach, recognized the existence of a reality beyond what appears to us, he made it clear that the objects of science are not the "things in themselves" that lie behind "the appearances," but rather the appearances themselves. This was a doctrine rejected by both Einstein and Gödel. Gödel made his objections explicit: Whereas it was a "fruitful viewpoint [to make] a distinction between subjective and objective elements in our knowledge (which is so impressively suggested by Kant's comparison with the Copernican system), [when such a doctrine] appears in the history of science, there is at once a tendency to exaggerate it into a boundless subjectivism. . . . Kant's doctrine of the unknowability of the things in themselves is one example. . . ."

Gödel, however, was not through with Kant. In an essay written in 1961 but never published, he noted that it was "a general feature of Kant's assertions that literally understood they are false, but in a broader sense contain deeper truths." He had in mind Kant's doctrine that in proving geometrical theorems we always need new geometrical intuitions. This, Gödel pointed out, is provably false. But if we substitute "mathematical" for "geometrical," the result is a truth that follows directly from Gödel's incompleteness theorem. What was needed, then, for the continual development of mathematics (and, one might add, philosophy), was "a procedure that should produce in us a new state of consciousness in which we describe in detail the basic concepts we use in our thought, or grasp other basic concepts hitherto unknown to us." This he claimed to have found in the later "transcendental phenomenology" of Edmund Husserl. "Transcendental phenomenology," he wrote in a draft of a letter to the mathematician-philosopher Gian-Carlo Rota, "carried through, would be nothing more nor less than Kant's critique of pure reason transformed into an exact science," which "far from destroying traditional metaphysics . . . would rather prove a solid foundation for it." In Husserl, Gödel thought he had found a form of idealism that, though derived from Kant's, was not incompatible with realism. That Husserl shared Gödel's disdain for unreconstructed Kantianism is apparent from a remark he made in 1915: "German idealism has always made me want to throw up."

Einstein's objections, in turn, to the new quantum mechanics—in particular his formulation of the EPR paradox—reflected a rejection of the Kantian turn in epistemology in its simplified reconstruction by positivists like Mach. The uncertainty principle, after all, is an example of the same tendency to draw ontological conclusions from epistemological premises, in this instance, from our inability in principle to know simultaneously the position and velocity of a subatomic particle, to the nonexistence of such a combined state. Not only did Einstein reject this reasoning, he resisted what he took to be Heisenberg's more fundamental belief that we should abandon the very idea of "quantum reality." For Einstein, as for Gödel, philosophy without ontology was an illusion, and physics without philosophy reduced to engineering. And for Einstein, engineering was a poor substitute for physics. When his eldest son, Hans, decided on an engineering career, his father wrote that he was pleased that Hans had found a subject to concentrate on but also that "what he is interested in isn't really important, even if it is, alas, engineering. One cannot expect one's children to inherit a mind."

An Offer That Couldn't Be Refused

Before Princeton, Einstein's and Gödel's philosophical sentiments had proceeded on parallel but independent lines. At the institute, however, the lines began to converge. In thought as in life, Einstein found himself increasingly entwined with Gödel. In 1949 Gödel received from his friend an unexpected housewarming gift, "a wonderful flower vase." At the time Einstein was celebrating his seventieth birthday, and "after long searches" Gödel finally settled on a birthday gift for his friend, an etching. (Adele had knit Einstein a sweater but decided against sending it.)

There was another gift as well. Gödel had been invited by P.A. Schilpp to contribute to a new volume in his series, "The Library of Living Philosophers," a volume in honor of Einstein's seventieth birthday, to be entitled *Albert Einstein: Philosopher-Scientist*. It is the only volume in the series to be devoted to a scientist. No doubt aware of Gödel's friendship with Einstein, Schilpp must have assumed his invitation was an offer Gödel couldn't refuse. He didn't. Gödel's essay would be his second exercise in this venue, following the 1944 essay on Russell. In time he would draft a third essay, this time on Rudolph Carnap, though he never submitted a final version, and he declined a fourth invitation to write on Popper.

Gödel wasted no time setting to work. Schilpp suggested a title, "The Realistic Standpoint in Physics and Mathematics," but Gödel rejected it. He had in mind an ontological investigation—the grand philosophical quest for the reality of time—reinterpreted as an examination of what relativity theory has to teach us about this question, which has exercised the philosophical imagination from Parmenides and Plato to Kant. Gödel informed Schilpp that he would submit a brief essay, of three to five pages, on the topic, "The Theory of Relativity and Kant." It is not known whether Schilpp appreciated the perversity of Gödel's proposal, in which he would put forward the thesis that Kant had anticipated Einstein. For Einstein was (and is) widely viewed not as having confirmed Kant but rather as refuting him. Kant dedicated a large part of his *Critique of Pure Reason* to the attempt to establish a priori that Newton was the final truth about physics and Euclid the last word on geometry, whereas Einstein demonstrated empirically that both Euclid and Newton were wrong. (Kant also claimed that logic would never take a step beyond Aristotle, a view made nonsense by Frege and Gödel. The great philosopher, it seems, got it completely wrong about Newton, Euclid and Aristotle. No one's perfect.)

What on earth was Gödel thinking? One thing is clear. It was not simply the long walks with Einstein that had aroused Gödel's interest in the problem of time. On his first visit to the institute, he had been delighted to attend a seminar on quantum mechanics given by his friend Von Neumann. He had begun his studies at the University of Vienna in physics and had maintained an active interest ever since. ("Active interest" in Gödel's case means a level of competence that for any normal person would constitute a career.) Since the question of time is at the center of special relativity, he could not have failed to attend to it. The primary source of his interest in time, however, was his preoccupation with "idealistic" philosophers, from Parmenides and Plato to Leibniz and Kant. Each of these thinkers, in his own way, questioned the ultimate reality of time. For Plato, things in time never really are but are always coming-to-be. Time itself, he said in his cosmological dialogue the *Timaeus*, is but a "moving image of eternity." Gödel was deeply sympathetic with Plato's philosophy, but his true hero was Leibniz. Why then did he choose to discuss, for the Schilpp volume, the relationship of relativity theory not to Leibniz but to Kant, for whom he had mixed feelings?

A partial answer is that in composing an essay on Einstein, Gödel was operating from within the modern scientific framework inaugurated by Newton. It is true that Gödel believed that physics in the modern era should have followed Leibniz, a philosopher in the tradition of German idealism, rather than his rival, Newton, but the fact remained that Einstein's framework was Newtonian. Consistent, then, with his methodology of undermining from within, Gödel chose to demonstrate certain (radical) formal and philosophical consequences from within the "host framework," modern physics from Newton to Einstein. As with his incompleteness results, he believed that this strategy made his formal derivations inescapable and his philosophical conclusions hard to resist.

A consequence of Gödel's methodology is that the question of "realism" in his Einstein essay becomes delicate. In that essay, he defends the reality of relativistic space-time, much as he promoted, in the philosophical implications he drew from his incompleteness theorem, the reality of the natural numbers, and, in his consistency proof for the continuum hypothesis relative to set theory, the reality of sets. But whereas he remained convinced of the fundamental reality of numbers from within his own philosophical system, not just that of its "host" framework, his preference for Leibniz leaves open the question of whether, from within his own (never fully realized) philosophy of space and time, he believed that the physical world as such was founded on something yet more fundamental, like the "monads" introduced by Leibniz. In his discussion of Einstein, however, he is operating within the philosophical framework of relativity theory, and so he adopts the familiar standpoint that space-time is part of the ultimate furniture of reality. The question he poses, then, is precisely this: if we believe in the ultimate reality of relativistic space-time, are we forced to be idealists about time?

Ontology and Epistemology: The Two Axes of Philosophy

. . . Almost all accounts of the concept of mathematical truth can be identified with serving one or another of these masters [semantics and epistemology] at the expense of the other.

PAUL BENACERRAF

The importance Gödel attached to his investigation of this question in his contribution to the Einstein volume cannot be overstated. A month after beginning his work in earnest, he wrote his mother that he was so preoccupied with it, he would have to give up writing letters until it was completed. Time, he told Wang years later, remains, even after Einstein, *the* philosophical question. Contrast this with what the influential ("analytical") philosopher of science Hilary Putnam has written: "I do not believe that there are any longer any philosophical problems about Time; there is only the physical problem of determining the exact physical geometry of the four dimensional continuum we inhabit." And Putnam is hardly alone. For Gödel, however, time is "that mysterious and seemingly self-contradictory being which, on the other hand, seems to form the basis of the world's and our own existence." Plato could not have put it better. Gödel's preoccupation with the question of the reality of time went against the grain of the philosophical tradition that dominated (and still dominates) his adopted country. His project more closely resembled that of "continental" philosophers like Martin Heidegger, whose magnum opus gives away the central problem in its title, *Being and Time*, and, more closely still, *On the Phenomenology of the Consciousness of Internal Time*, by Edmund Husserl, a work to which Gödel would in the course of time devote a great deal of attention, with special reference to Husserl's distinction between physical time and "internal time-consciousness." Unlike Heidegger, however, as well as other philosophers of time such as Henri Bergson and J.M.E. McTaggart, who, like Gödel, took the question of the ontological implications of the reality of time seriously, Gödel had no interest in yet another inconclusive metaphysical debate in which the elusiveness of the subject would be matched only by the density of the surrounding prose. He chose instead to examine the traditional philosophical question from within an untraditional mathematical context, Einstein's theory of relativity. The idea, as with his incompleteness theorem, was to establish formal results that would have deep philosophical implications. Gödel would be a pathfinder, then, along two directions: a mathematical approach to the philosophy of time, and a philosophical assessment of the mathematics of relativity. Einstein himself had been reluctant to engage in the latter. The few remarks he did allow himself verged on inconsistency. He agreed with the philosopher Emile Meyerson that the temporal component of space-time was not a mere fourth "spatial" dimension, but at other times he spoke of the "rigid four-dimensional space of the special theory of relativity." When his friend Michele Besso died, he wrote to his widow that "now he has preceded me a little by parting from this strange world. This means nothing. To us believing physicists the distinction between past, present, and future [i.e., between what is now and what is not now] has only the significance of a stubborn illusion." Yet Carnap wrote that Einstein had told him that "the now means something special for man, something which physics cannot speak to." His physics was already deeply philosophical. He had no wish to take time away from scientific research to match the philosophers in discussions about the nature of existence.

Gödel did. He would complete the philosophical journey Einstein had begun in the theory of relativity. By stretching Einstein's mathematics to the limit, he would simplify and clarify the philosophy, showing the world that one can plumb the depths of being and time without disappearing into a Black Forest of Heideggerian prose. To appreciate how Gödel was able to accomplish this it is necessary to understand that the two fundamental axes along which the course of philosophy is plotted are ontology and epistemology.

You can assess any position in philosophy by the relationship it proposes between being and knowing. Some traditions, like the Greek one of Plato and Aristotle, place ontology in the center, while others, like the modern period inaugurated by Descartes, put the emphasis on epistemology. Clearly, however, a complete philosophy will have to do justice to both. Unfortunately, there is a deep and irreducible tension between the two perspectives that makes a reconciliation difficult to achieve. The ontological perspective, "the view from nowhere" (to exploit Thomas Nagel's evocative phrase) seems to leave no room for "the world as I found it" (to borrow an expression from Wittgenstein). If we note further that the world as we experience it, "our world," is essentially temporal, whereas the logical and empirical conditions for existence as such are more spatial, in the most general sense, than temporal, we begin to take a true measure of the problem Gödel set for himself. We also begin to appreciate the sympathy Gödel would come to feel for Husserl's later phenomenology, the goal of which was to do justice to ontology without neglecting epistemology, a goal attempted by Kant, whose system, nevertheless, by Gödel's lights, failed to do justice to one element of the dialectic: ontology.

Now, Gödel understood that the advent of relativity theory enabled one for the first time to cast the question of the reality of time into a theoretical context amenable to formal mathematical methods. His approach to the philosophy of time, then, would take the form of a frontal assault on the ontological implications of relativity theory. Can one consistently maintain both the existence of time, intuitively understood, and the truth of relativity theory? This was a peculiar question, however, because special relativity was an epistemologically inspired theory (which is what made it the apple of the positivists' eye) that sprang from such a priori assumptions as that time is determined by the measurement of simultaneity via synchronized clocks, and such a posteriori observations as that different clocks in different inertial frames deliver different simultaneity results and that there is no objective method for privileging one inertial frame above all others. Einstein himself, then, had already drawn ontological conclusions from epistemological premises. Gödel continued drawing conclusions beyond the point where Einstein stopped. He would out-Einstein Einstein by taking the physicist's own ontological reasoning to its logical conclusion. And he would do this by answering a question that Einstein, though fully aware of it, wished to sidestep: is the temporal component t of four-dimensional relativistic space-time, i.e., what remains of "time" after relativity theory, really time?

In his response to Gödel's paper in the Schilpp volume, Einstein acknowledged that "the problem here involved disturbed me at the time of the building up of the general theory of relativity." This problem he described as follows: "Is what remains of temporal connection between world-points in the theory of relativity an asymmetrical relation [like time, intuitively understood, and unlike space], or would one be just as much justified ... to assert A is before P [as to assert that A is after P] ... ?" The issue could also be put this way: is relativistic space-time in essence a space or a time?

An Atom Smasher for the Mind

Even to raise this question, one must first distinguish between time in-tuitively understood and t, the temporal component of relativistic space-time. While Einstein himself was attuned to this distinction, the very success of relativity theory had encouraged most thinkers to con-flate the two concepts. The situation was deeply reminiscent of what Gödel had encountered when he constructed his incompleteness theo-rem in response to the Hilbert program, where the issue was the rela-tionship of formal demonstrability to intuitive mathematical truth. Here too, many researchers conflated the two, making the idea of a "relationship" seem moot. Let us recall Gödel's comment to Wang: "... formalists considered formal demonstrability to be an analysis of the concept of mathematical truth and, therefore, were of course not in a position to distinguish the two." With equal justification, he could have said that relativistic physicists and analytical philosophers of sci-ence were not in a position to distinguish the temporal component of four-dimensional relativistic space-time from the intuitive concept of time. Gödel's essay on Einstein, then, was not an "excursion," as it is often taken to be, but rather a continuation of the "Gödel program" of testing the limits of formal methods in capturing intuitive concepts.

Having raised this issue, how does one then devise a thought ex-periment to distinguish the two mathematically (if indeed they are dis-tinguishable)? In subatomic physics, one can submit particles to the extreme forces of an accelerator, or "atom smasher," which results in particles that are indistinguishable under less-extreme conditions re-vealing themselves as distinct. Gödel would devise a method for subjecting the concept of space-time to similarly extreme conditions—in this case, geometrically, not dynamically, extreme—so that invisible differences between the two concepts would become manifest. This too was a continuation of a methodology Gödel had employed in his incompleteness theorem. The method consists in creating what can be called *limit cases*, formal constructions that by design are so extreme that they limit, mathematically, the possible intuitive interpretations they will admit. For his incompleteness theorem, Gödel devised a formal system together with a series of ingenious definitions and coordinations for which it could be demonstrated that the concept of formal proof, as it appeared in the system, could not, on pain of contradiction, be interpreted as representing intuitive mathematical truth. He did this by constructing a formula that was provably unprovable, but intuitively true.

In his contribution to the Einstein volume, Gödel would construct a world model for the equations of general relativity whose geometry was so extreme that the temporal component of the resulting space-time structure could not reasonably be seen as representing intuitive time. Einstein had already succeeded, in the theory of relativity, in bringing about the geometrization of physics. What Gödel did was to construct a limit case for the relativistic geometrization of time. He would do this by bringing to the fore various properties that anything deserving the name of intuitive time would have to possess, including Einstein's requirement that the series of events be asymmetrically ordered, so that if A is before B, it cannot also be after B. Gödel would then demonstrate mathematically that in the world model he had constructed, there were continuous timelike world lines connecting any two events, so that even if B were observed occurring after A, one could undertake a journey—in a very fast spaceship—that would take one to B before one reached A. From this, Gödel would conclude that the space-time structure in such a world was clearly a space, not a time, and therefore that t, the temporal component of space-time, was in fact another spatial dimension—not time as we understand it in ordinary experience.

A journey along the closed, continuous timelike world lines Gödel had discovered in (what came to be known as) the Gödel universe could only be described as *time travel*. Gödel had achieved an amazing demonstration that time travel, strictly understood, was consistent with the theory of relativity. Enthusiasts of time travel would in due course become excited by this discovery, but they would fail to see that the primary result was a powerful argument that if time travel is possible, time itself is not. If his results held up and his interpretation of them survived scrutiny, Gödel would have succeeded in demonstrating, mathematically, a result about the reality (or unreality) of time that had eluded idealist philosophers for centuries, from Plato to Kant, and he would have done so, once again, as a spy, this time in the house of physics. Before Einstein's very eyes, a metamorphosis had occurred. The theory he had devised to capture time, to pin it down mathematically and render it amenable to human understanding, had been transformed, in Gödel's hands, into a disappearing hat trick.

Einstein was impressed. "Kurt Gödel's essay," he wrote, "constitutes, in my opinion, an important contribution to the general theory of relativity, especially to the analysis of the concept of time." But he set the stage for how others would respond when he stated that "it will be interesting to weigh whether these [cosmological solutions] are not to be excluded on physical grounds." Most thinkers, once they had recovered from the shock of Gödel's discovery, would restrict their response to enquiring whether the Gödel universe was sufficiently realistic from a physical standpoint to be taken seriously.

Einstein himself, however, had a bad track record in acknowledging mathematical consequences of general relativity as physically realistic. When Karl Schwarzschild, a German colleague, discovered in 1916 that if a star began an extreme gravitational collapse into itself, its mass would eventually reach a critical point after which space-time would be so severely curved that nothing inside (what is now known as) the "event horizon," including light, would be able to escape, Einstein dismissed the "Schwarzschild singularity" as a mathematical anomaly with no physical significance. He was wrong. We now call these singularities "black holes," and astronomers find them at the center of every galaxy.

Later, in 1917, the Dutch astronomer Wellem de Sitter proposed a cosmological model for general relativity in which the universe was not static, as Einstein believed it to be, but rather expanding. Still later, in 1922, Aleksandr Friedmann, a Russian mathematician and physicist, argued that it was a consequence of general relativity that the universe must be either contracting or expanding. Einstein rejected both de Sitter and Friedman as having produced unphysical mathematical models of the universe. But since their ideas were consistent with general relativity, he proposed a new, ad hoc principle to be added to relativity, the "cosmological constant," whose sole purpose was to introduce an antigravitational law that would counteract the forces of gravity that de Sitter and Friedman had deduced would otherwise cause the universe to spiral inwards or expand outwards. Once again, Einstein turned out to be wrong. We now have overwhelming empirical evidence that the universe is indeed expanding, and Einstein would eventually call the introduction of the cosmological constant "my greatest blunder." Yet here too he would be wrong: there now appears to be an antigravitational force that is accelerating the universe's expansion.

In all of these cases, Einstein rejected world models for general relativity on the grounds that the extreme geometrical conditions they represented were inconsistent not with the letter of general relativity, but with his own intuitions about the shape of the universe. Each time, however, the real world refused to cooperate with the great physicist's a priori demands. It should not, then, be taken as decisive that when Gödel proposed a new, geometrically extreme world model for general relativity, Einstein was inclined to question its physical significance.

Yet Einstein did take Gödel to have made an important discovery about the nature of time. The question was what exactly this discovery meant. What had Gödel really been up to in this beautiful and mysterious work? On this there was a deafening silence. ("Like most others," wrote a distinguished philosopher of physics decades after Gödel's essay first appeared, "I avoid the puzzling issue of what Gödel really thought he was showing about time and stick to the easier stuff on closed timelike loops.") Unlike Gödel's achievements in logic, which, after the initial shock wore off, became gradually understood, his cosmological results on Einstein's theory of relativity remained an enigma, and all too soon, a distant memory. The question, however, cannot be avoided: What really happened when Gödel became Einstein? How exactly did he achieve his shocking results on time and relativity, and what should physicists and philosophers have made of them? After all is said and done, can it really be true that we, who look forward to dessert while nibbling on our salads, are living in a world without time?

7 | The Scandal of Big "T" and Little "t"

"Scientific people," proceeded the Time Traveler, "... know very well that time is only a kind of space."

H.G. WELLS, THE TIME MACHINE

Albert Einstein: Philosopher-Scientist, dedicated by P.A. Schilpp to the physicist on the occasion of his seventieth birthday, was a great success. It remains the most influential volume in the Library of Living Philosophers, not least because of its wonderful debate between Einstein and Niels Bohr, friends and adversaries, on the future of the Copenhagen interpretation of quantum mechanics. Einstein's walking companion, in turn, had worked intensely on his own contribution, writing to his mother that the work left him little leisure for correspondence. That summer, he canceled as well his accustomed vacation trip to the seashore. When the auspicious volume finally appeared, he cannot have failed to be disappointed by the near silence with which his essay was greeted.

To be sure, there was a minor stir among astrophysicists and cosmologists concerning the validity of Gödel's construction of new world models for general relativity. The initial response, however, was that he had simply made a mistake in his physics. No less a physicist than S. Chandrasekhar, who had attended a talk Gödel had given on his new models at Princeton, published an article with J.P. Wright in the *Proceedings of the National Academy of Sciences* claiming that Gödel had made an error when he described the possibility of time travel as along a geodesic—the path of inertial motion, or free fall—in the Gödel universe. This no doubt contributed to the lack of interest in Gödel's essay among philosophers. If the physical premise was faulty, why bother to examine the philosophy?

But was Gödel really in error? Amazingly, the editors of the *Proceedings* had not seen fit to consult the author himself before publishing a report of his alleged error concerning an elementary concept of relativity theory. Might it not have been Chandrasekhar and Wright, not Gödel, who had made a mistake? This possibility seems not to have occurred to the editors, yet it turned out to have been the case, a fact demonstrated not by a physicist but by a philosopher, Howard Stein, who showed clearly that time travel in the Gödel universe could take place only under great acceleration, which could be provided by a spaceship, not along the free-fall path of a geodesic. More astonishing yet, however, Stein could not get the correction of Chandrasekhar and Wright accepted for publication. Only when Gödel himself intervened did the fact finally make it into print that his argument for the possibility of time travel was relativistically valid.

What had gone wrong? Clearly, regardless of Gödel's reputation as a great logician, the astrophysics community saw him as an outsider, and moreover as attempting to swim against the intellectual tide. But the scandal of disregard extended to philosophy as well. Gödel's contribution to the Schilpp volume had almost no impact on the community of philosophers. Except for a few highly technical discussions of the physics, with some brief though poignant glances at Gödel's philosophical goals, his argument that relativity theory, correctly understood, provides strong support for the great philosophers throughout history who were skeptical of the objective reality of time, went unheeded. Naturally, there was some interest in the question of time travel. There always is. It is a topic of perennial fascination among thoughtful and imaginative people, and the fact that Gödel had derived such an exotic conclusion from the respectable equations of relativity inevitably raised a few eyebrows. But on the question of whether he had succeeded in showing that time is ideal there was a profound silence. If Gödel had not been taken seriously as a philosopher before his contribution to the Schilpp volume, nothing changed after its appearance.

Quite simply, he had never been a member of the club: he was out of touch and out of step with the philosophical establishment, in Princeton as elsewhere, and the reason was not hard to fathom. Just as Wittgenstein's language-centered early work, the *Tractatus*, had helped set the philosophical agenda following World War I, not least in Gödel's Vienna, in the aftermath of the Second World War, Wittgenstein's later, still linguistically oriented philosophy came to dominate again, this time in Gödel's newly adopted country. To many philosophers, it must have seemed as if Gödel had slept through not one but two Wittgenstein revolutions. It added insult to injury that W.V.O. Quine, the dominant figure for years in American philosophy and the most analytic of analytical philosophers, was also absent from Gödel's thinking. Gödel himself was acutely aware of this alienation. When the time came for his essay on Cantor's continuum problem to be reprinted in a now classic collection coedited by his Princeton colleague Paul Benacerraf and Harvard's Hilary Putnam, he would not agree to the republication until he had been convinced that the editors would not deride his essay. Shameful it may have been that coming out of nowhere in every sense his highly compressed, paradoxical-sounding philosophical defense of temporal idealism, based on an arcane new cosmological model for an abstruse physical theory, failed to arouse more than a murmur. But it was not surprising. Yet why, more than half a century after it was proposed and ignored, is Gödel's argument still worthy of attention? What had Gödel really accomplished?

What Your Parents Never Told You About the Age of the Universe

The problem Gödel inherited from Einstein had been understood for centuries to concern the most fundamental aspect of human experience.

For Kant, space and time are the two essential "forms of human sensibility," with time, as the form of both "inner" and "outer sense," being the more basic. Yet time is far more elusive than space. Capturing time through mathematics—a form of thought from which philosophers since Plato have taken pains to remove anything remotely temporal—is like trying to trap water with a net. With the advent of Einstein's theory of relativity, however, the mystery of this form of being was widely taken to have been resolved. Philosophers could finally relax. Einstein had taken care of business.

Appearances, however, can be deceptive. The universe, for example, as everyone knows, is very old. Its exact age is a matter for debate, but there is no disagreement that it runs to billions of years. We marvel that as frail and isolated a species as we are can have achieved such impressive wisdom about the origins of everything that is. In truth, however, it is more than marvelous to have discovered the age of the universe. It is impossible. For if the universe is n years old, its present state comes n years after the moment when it all began. In 1905, however, Einstein had demonstrated in the special theory of relativity that there is no such thing as "the present state of the universe," that is, what would be revealed by a snapshot of the universe as it exists at this very moment. The relativity of simultaneity implies that what is taken to be "now" relative to one inertial frame will differ from what is "now" in another frame if the second frame is in motion relative to the first. It follows immediately that if the theory of relativity is correct, there simply is no such thing as "the present state of the entire universe" of four-dimensional space-time. Einstein himself said this quite clearly: "The four-dimensional continuum is now no longer resolvable objectively into sections, all of which contain simultaneous events; 'now' loses for the spatially extended world its objective meaning."

None of this was lost on Gödel. To him, there was an inconsistency between Einstein's theory and the everyday belief that time, unlike space, "passes" or "flows." On this question, two assumptions dominated, then as now, in the popular as well as the scientific consciousness. Both of them are faulty. The first is that special relativity is compatible with the passing of time, as long as it is acknowledged that this flow has only local, as opposed to global, significance. The other is that the world according to special relativity is a fixed four-dimensional space-time "block," but that this does not conflict with the deliverances of ordinary experience. The former fails because whatever the flow of time is, it is not a merely geometrical fact and thus cannot enjoy only local existence. A river's course, for example, may curve locally—may be serpentine near us but straight elsewhere—but what would it mean to say that the river flowed only in our neighborhood?

Concerning the second assumption, one need only recognize the befuddlement that would ensue if one were to try to act on the assumption that today's breakfast is no more actual than yesterday's or tomorrow's, that the future, like the present, has already arrived. ("The future is now," reads the logo of Hudsucker Enterprises in the film *The Hudsucker Proxy*. This makes for an entertaining story but an unconvincing metaphysics of everyday life.) Should I still be wondering what to order for breakfast yesterday, as I am for tomorrow, or should I cancel both orders because the meals have already arrived? And since the present is no more real than the past and I am still lying on the beach as I was last summer, why am I identifying only with the "I" that is presently shivering in the cold? Am I simply making a mistake? Or are there as many "I's" as there are moments in time, and if so, are they all me, or only parts of me? (I have spatial parts, of course: head, hands, feet; do I also have "temporal parts"?) The confidence of the popular (and not so popular) mind is misplaced when it clings to the belief that all is well, temporally speaking, between the universe and Dr. Einstein. All is not well at all.

But as indicated by its name, special relativity is not the full theory of relativity. Its validity is restricted to so-called inertial reference frames, those that are unaccelerated and move in straight lines. The final, comprehensive theory, general relativity, has no such restrictions. It includes an account of gravity, the first theory of gravity to replace Newton's. Since it is gravity that governs the universe as a whole, general relativity is the foundation of the modern science of cosmology. If special relativity, moreover, introduced the discovery that matter is equivalent to energy, the general theory announced the identity of gravity with space-time curvature. Matter in motion determines the shape of space-time. The possibility arises that some reference frames might be privileged, namely those that follow, as Gödel put it, the mean motion of matter in the universe. Time relative to those frames of reference bears the designation "cosmic time," and this opens up the possibility that time in something like the pretheoretical sense might after all be consistent with relativity, in particular with general relativity. It is time in this sense that is (or should be) invoked when cosmologists speak of the age of the universe.

The question remains, however, how closely this new concept of time resembles what time was thought to be before Einstein. The astrophysicist James Jeans, whom Gödel would cite by name when he came to discuss these issues, thought the resemblance was very close indeed. With the advent of general relativity and cosmic time, "time regained a real objective existence, although only on the astronomical scale." Since, moreover, every known relativistically possible universe "makes [in this way] a real distinction between space and time," Jeans believed, "[we have] every justification for reverting to our old intuitional belief that past, present, and future have real objective meaning." In short, "we are free to believe that time is real." Just this Gödel would put to the test.

What Gödel Means by Time

At issue is the leitmotif of Gödel's lifework, the dialectic of the formal and the intuitive, here, of formal versus intuitive time, between what remains of time in the theory of relativity and the time of everyday life. The difference between these two conceptions is crucial. It can be illuminated by considering what the early-twentieth-century philosopher J.M.E. McTaggart called the A-series and B-series. The B-series is founded on the characterization of dates and times in terms of the fixed relationship of "before" and "after." It is a structurally or "geometrically" defined series, analogous to a space. It is the temporal series captured by calendars and by history books. The year 1865, for example, comes—now and forever—before 1965 and after 1765, and these structural, "geometric" facts are fixed and unchangeable. The A-series, in contrast, is essentially fluid or dynamic. It contains the "moving now," i.e., the present moment, which is always in flux. That your dentist appointment is at 3 p.m. on May 19 is a B-series fact that has been marked on your calendar for months. It will remain a fact after the appointment is long forgotten. That now, however, is the very date and time of the appointment is a scary A-series fact that has not obtained until this very moment, and will happily no longer obtain tomorrow. (It is no accident that a famous philosophical essay on the A-series is entitled "Thank Goodness That's Over.")

Though the A-series represents, intuitively, the most fundamental aspect of time—indeed, what distinguishes time from space—it is marked by several concomitants, each one difficult to capture in the formal language of mathematics. First is the fact that one time—now—is privileged over all others. This privilege passes from time to time. What is now will soon be then. Second, according to this conception, time passes, or flows, or lapses, and in a certain "direction": what is future becomes present, then past. Third, unlike both space and the B-series, "position" in the A-series is not ontologically neutral. Whereas to exist in New Jersey is to exist no less than in New York (protests by New Yorkers notwithstanding), to "exist in the past" is no longer to exist at all. Socrates had his time on stage, but it passed, he died, and his name has been removed from the rolls. (It follows that there is nothing subjective or mind-bound about the A-series, i.e., about what is happening now. If there is such a thing as "inner time"—the subject, it would appear, of Husserl's investigations—then this must be distinguished from the A-series.) Fourth, while the past has passed and is now forever fixed and determinate, the future remains, as of now, open. Simultaneity, finally, since it determines what really exists at the same time as other things exist, is absolute and non-relative. We cannot, merely by choosing a frame of reference, determine what really exists at this moment. Either my friend in Paris is speaking on the phone at very same time at which I am writing this, or she isn't, regardless of how I try to determine, via synchronized clocks, whether her speaking is occurring at the same moment as my writing.

Intuitively, time is characterized by both the A- and the B-series. If time as we experience it in everyday life, however, is to be identified with formal time—time as it is studied in physics—a problem arises. What we call "t," the temporal component of relativistic space-time, can be consistently interpreted as representing the B-series. The problem lies with the A-series. Since, as Einstein put it, in special relativity "'now' loses for the spatially extended world its objective meaning"—that is, there is no objective, worldwide "now"—it appears that "t" cannot represent the A-series, in which there is a single worldwide "now" whose "flux" constitutes the change in what exists that characterizes temporal, but not spatial, reality. This should come as no surprise. One of the most striking characteristics of relativistic space-time is that space and time are no longer to be considered independent beings but rather two inextricably intertwined components of a single new kind of being, not space or time but rather space-time.

The A-series cannot be made to resemble space. What keeps this seemingly obvious fact hidden from many formal thinkers, whether physicists or logicians, is that in special relativity, "t" is formally distinguished from the three spatial dimensions. In the definition, for example, of the space-time "interval"—the unique relationship between any two space-time events that is frame-invariant, hence agreed upon by all observers, no matter their state of motion—the temporal variable, "t," is distinguished from the three spatial variables by being preceded by a negative sign. All this demonstrates, however, is that time in special relativity has a different "geometry" from the spatial dimensions, not that it is a qualitatively different kind of being, namely something that "flows." To be blind to this fact is to confuse the formal with the intuitive.

It is not for nothing that with the theory of relativity Einstein is said to have accomplished the geometrization of physics (an achievement for which, as we have seen, he owed a great debt to the mathematician Minkowski, his long-suffering teacher at the Technical Institute in Zurich, who took the bold step of re-creating special relativity in a four-dimensional geometric framework). It is not just that Einstein reconceived the geometry of the universe. Rather, in special relativity, he made the defining characteristic of time not its qualitative distinction from space, as Kant and Newton had done, but rather its contribution to the geometry of four-dimensional space-time. Similarly, in general relativity, he not only provided a new geometry for the laws of gravity, he defined gravity itself geometrically, as space-time curvature. One of Einstein's claims to fame, after all, is his uncanny ability not only to provide new descriptions of old phenomena but new definitions as well. In this, as in many other aspects of his discoveries, he is as much philosopher as physicist. The coup de grâce came when he replaced Newton's intuitively evident Euclidean mathematics with unintuitive non-Euclidean geometry.

Time as it appears in relativity theory, then, was ripe for consideration in the "Gödel program" of assessing the extent to which intuitive ideas can be captured by formal concepts. This is what Gödel had in mind when he titled his contribution to the Schilpp volume, "A Remark About the Relationship Between Relativity Theory and Idealistic Philosophy." The "idealistic philosophers" he was referring to were thinkers like Parmenides, Plato and Kant, who questioned whether our subjective experience of the flow of time has an objective correlative. To such thinkers, time was always an ontological suspect. As before, when he examined the relationship of intuitive arithmetic truth, or big "T," to its representation as formal mathematical proof in Russell's *Principia Mathematica*, Gödel would begin by clarifying the distinction between intuitive time and little "t," its formal representation in Einstein's theory of relativity as the temporal component of four-dimensional Einstein–Minkowski space-time. Drawing from his contribution to the Schilpp volume as well as the longer versions of this essay that have now been published, we can say that Gödel characterized intuitive time—"what everyone understood by time before relativity theory"—as "Kantian," or "prerelativistic." Time in this intuitive sense, he said, is "a one-dimensional manifold that provides a complete linear ordering of all events in nature." This "objective lapse of time" is "directly experienced" and "involves a change in the existing [i.e., in what actually exists]." Time in the intuitive sense, for Gödel, is something "whose essence is that only the present really exists." In particular, it "means (or is equivalent to the fact) that reality consists of an infinity of layers of 'now' which come into existence successively." These features Gödel took to be essential properties of time in the intuitive sense, since "something without these properties can hardly be called time." Clearly, time so characterized is reflected in the A-series, and indeed Gödel refers to McTaggart by name in his essay. The question that remains is whether this intuitive concept can be captured by the formal methods of relativity.

Gödel's Dialectical Dance with Time

As he had previously done in his incompleteness theorem, Gödel demonstrated that those who fail to grasp the distinction between the intuitive and the formal concept are not in a position to make a proper assessment of their relationship. Having made that distinction with remarkable clarity, he was able to establish, by an ingenious and entirely unsuspected formal argument—which in itself, as Einstein pointed out, was a major contribution to relativity theory—the inability of the formal representation to capture the intuitive concept. Gödel's dialectical dance with intuitive and formal time in the theory of relativity contained an intricate series of steps. We begin with a large-scale view of the structure of Gödel's argument, then move on to a closer examination. First the forest, then the trees.

The opening move concerns the more limited special theory of relativity. Given that the A-series contains the flux of "now," the absence of an objective, worldwide "now" in special relativity rules out its existence. But absent the A-series there is no intuitive time. What remains, formal time as represented by the little "t" of Einstein–Minkowski space-time, cannot be identified with the intuitive time of everyday experience. The conclusion, for Gödel, is inescapable: if relativity theory is valid, intuitive time disappears.

Step two takes place when Gödel reminds us that special relativity is "special" in that it recognizes only inertial frames in constant velocity relative to each other. It does not include an account of gravity. Einstein's general theory of relativity, in contrast, of which the special is a special case, does. In general relativity, as we have seen, gravity itself is defined as space-time curvature, determined, in turn, by the distribution of matter in motion. It follows that whereas in special relativity no frames of reference or systems in motion are privileged, in the general theory some are distinguished, namely those that, in Gödel's words, "follow the mean motion of matter" in the universe. In the actual world, it turns out, these privileged frames of reference can be coordinated so that they determine an objective remnant of time: the "cosmic time" we encountered earlier. In general relativity, then, time (of a sort) reappears.

But no sooner has time reentered the scene than Gödel proceeds to step three, where he exploits the fact that Einstein has fully geometrized space-time. The equations of general relativity permit alternative solutions, each of which determines a possible universe, a relativistically possible world. Solutions to these complex equations are rare, but in no time at all Gödel discovers a relativistically possible universe (actually, a set of them)—now known as the Gödel universe—in which the geometry of the world is so extreme that it contains space-time paths unthinkable in more familiar universes like our own. In one such Gödel universe, it is provable that there exist closed timelike curves such that if you travel fast enough, you can, though always heading toward your local future, arrive in the past. These closed loops or circular paths have a more familiar name: *time travel*. But if it is possible in such worlds, Gödel argues, to return to one's past, then what was past never *passed* at all. But a time that never truly passes cannot pass for real, intuitive time. The reality of time travel in the Gödel universe signals the unreality of time. Once again, time disappears.

But the dance is not over. For the Gödel universe, after all, is not the actual world, only a possible one. Can we really infer the nonexistence of time in this world from its absence from a merely possible universe? In a word, yes. Or so Gödel argues. Here he makes his final, his most subtle and elusive step, the one from the possible to the actual. This is a mode of reasoning close to Gödel's heart. His mathematical Platonism, which committed him to the existence of a realm of objects that are not accidental like you and me—who exist, but might not have—but necessary, implied immediately that if a mathematical object is so much as possible, it is necessary, hence actual. This is so because what necessarily exists cannot exist at all unless it exists in all possible worlds.

This same mode of reasoning, from the possible to the actual, occurs in the "ontological argument" for the existence of God employed by Saint Anselm, Descartes and Leibniz. According to this argument, one cannot consider God to be an accidental being—one that merely happens to exist—but rather a necessary one that, if it exists at all, exists in every possible world. It follows that if God is so much as possible, He is actual. This means that one cannot be an atheist unless one is a "superatheist," i.e., someone who denies not just that God exists but that He is possible. Experience teaches us that ordinary, garden-variety atheists are not always willing to go further and embrace superatheism. Following in the footsteps of Leibniz, Gödel, too, constructed an ontological argument for God. Then, concerned that he would be taken for a theist in an atheistic age, he never allowed it to be published.

In arguing from the mere possibility of the Gödel universe, in which time disappears, to the nonexistence of time in the actual world, Gödel was employing a mode of reasoning in which he had more confidence than most of his philosophical colleagues. In the case of the

Gödel universe, he reasoned that since this possible world is governed by the same physical laws that obtain in the actual world—differing from our world only in the large-scale distribution of matter and motion—it cannot be that whereas time fails to exist in that possible world, it is present in our own. To deny this, Gödel reasoned, would be to assert that "whether or not an objective lapse of time exists (i.e., whether or not a time in the ordinary sense exists) depends on the particular way in which matter and its motion are arranged in this world." Even though this would not lead to an outright contradiction, he argued, "nevertheless, a philosophical view leading to such consequences can hardly be considered as satisfactory." But it is provable that time fails to exist in the Gödel universe. It cannot, therefore, exist in our own. The final step is taken; the curtain comes down: time really does disappear.

Into the Forest

Such, in broad outlines, is the structure of Gödel's argument. Even from this brief sketch, it should be apparent how complex and subtle was the case Gödel made for the ideality of time, a far cry from the amateurish philosophical fumblings with which he is frequently credited. To appreciate the full force of his reasoning, however, it is necessary to look more closely at the details of his argument, to get close to the trees in the forest. His very first step, from little "t" and special relativity to temporal idealism, went unappreciated, in part because, as he remarked about his incompleteness theorem and big "T," mathematical truth, there was a widespread failure to appreciate the distinction between the formal and the intuitive. There still is. Even today, one can find distinguished proponents of the view that special relativity implies only that the flow of time must be tied to a frame of reference, and that the relativity of simultaneity—combined with the fact that the progress of "now" represents the flux of reality—simply means that reality itself must be relativized to a frame of reference.

The question not asked is this: does this conclusion make any sense? Fifty years ago Gödel had the answer: "the concept of existence . . . cannot be relativized without destroying its meaning completely."

How does Gödel know this? Perhaps relativity has revised what we mean by existence? This Gödel considered nonsense. Science, he maintained in his discussions with Hao Wang, does not analyze concepts, as does philosophy. It applies them. "The notion of existence," in particular, "is one of the primitive concepts with which we must begin as given. It is the clearest concept we have." To appreciate the force of Gödel's reductio ad absurdum, then, it is first necessary to recognize the absurdum. Not everything can be relativized. You can relativize velocity to a frame of reference. You can recognize that what's on my left is on your right and that what is here for me is there for you, that is, that when I say it is raining here, you agree that it is raining there. But reality as such is absolute. One cannot speak coherently of "my reality" or "your reality," "reality here" versus "reality there." When people say things like "my reality is a world in which people care for each other," they mean—or should mean—that this is their subjective view of the world, how it is or should be. But there is still only one objective reality, which includes the fact that this is your view of the world. If a doctrine implies the opposite, it is that doctrine that has to go. We can have a world in which there is time or a world in which there is existence, but not both. Gödel made the only rational choice: a world without time.

Since there is no single objective worldwide "now" in special relativity, and since there cannot be multiple rivers of time each of which determines the advance of reality, it follows that there simply is no such thing as the universal, worldwide flux of "now" or lapse of time consistent with relativity. As Gödel put it, "each observer has his own set of 'nows,' and none of these various systems of layers can claim the prerogative of representing the objective lapse of time." Special relativity, then, is not simply "incomplete" with respect to intuitive time. Einstein's theory *is inconsistent* with the existence of the A-series, with the reality of time in the intuitive sense. There is simply no way around it: if time as it is experienced in ordinary life is to be not ideal but fully real, Einstein must be wrong. And so he is. Or rather, the special theory of relativity is to be replaced by the general theory, which contains a universal theory of motion, including acceleration due to gravity. Hope remains. But this hope too Gödel will quash, beginning with the second step of his argument.

No Time for Time Travel

In general relativity, as we have seen, one can define, if not time itself, at least a kind of simulacrum of the real thing, namely, "cosmic time," determined by those frames of reference whose motion follows the mean motion of matter in the universe. This was a possibility opened up by Einstein's geometrization of space and time. The only constraints placed on this geometrization are those determined by the laws of general relativity. Any possible universe that obeys these rules must be, by the letter of relativity, physically possible. What Gödel discovered—by the judicious use of ingenious new geometrical methods that themselves constituted an important advance in relativistic mathematics—was that there are solutions to the equations of general relativity that provide world models in which all matter is rotating. Yet absent Newton's absolute space, with respect to what is the universe supposed to be rotating? "As a substitute for absolute space," said Gödel, "we have a certain inertial field which determines the motion of bodies upon which no forces act. . . . This inertial field determines the behavior of the axis of a completely free gyroscope." This is what Gödel used to define universal rotation: "It is with respect to the spatial directions defined in this way (by a free gyroscope . . .) that matter will have to rotate." (Gyroscopes, it will be recalled, entered Einstein's thought elsewhere, when he helped improve their design for use on U-boats during WWI.) In these rotating or "Gödel universes," Gödel proved, no single objective cosmic time can be defined. The last remnant of something even approximating intuitive time cannot be introduced into these Gödel universes, on pain of contradiction.

If one stake in the heart is good, two are better. Gödel discovered that in a subclass of the rotating universes, those that are not expanding, the large-scale geometry of the world is so warped that there exist space-time curves that bend back on themselves so far that they close; that is, they return to their starting point. A highly accelerated spaceship journey along such a closed path, or world line, could only be described as time travel. And it would be some spaceship. Gödel worked out the length and time for the journey, as well as the exact speed and fuel requirements. The top speed would be a significant fraction of the velocity of light, and the fuel requirements, too, would be enormous. (One theorist has calculated that even with a perfectly efficient rocket engine, the spaceship would require 10^{12} grams of fuel for every 2 grams of payload.) Paradoxically, however, the very fact that this inconceivably fast spaceship would return its passengers to the past demonstrated, by Gödel's lights, that time itself—hence speed and motion—is but an illusion. For if we can revisit the past, it still exists. How else could it be revisited? You can't revisit New Jersey if New Jersey is no longer there, and you can't return to time t if t has departed from the realm of existence. Thus temporal distance—past and future—turns out to be as ontologically neutral as the measure of space. This is something that even the "friends of Gödel," who in recent years have stepped forward to defend his account of time travel as logically and physically coherent, have failed to note. For Gödel, if there is time travel, there isn't time. The goal of the great logician was not to make room in physics for one's favorite episode of *Star Trek*, but rather to demonstrate that if one follows the logic of relativity further even than its father was willing to venture, the results will not just illuminate but eliminate the reality of time.

Protecting Time from Gödel

Such, in essence, was the argument put forward by Gödel in "A Remark About the Relationship Between Relativity Theory and Idealistic

Philosophy," a gift for his friend Albert Einstein in the Schilpp volume dedicated to the great physicist on the occasion of his seventieth birthday. Six pages was all Gödel needed to defeat time. Over fifty years later, however, what Gödel really accomplished in this brief compass remains hidden. He had once again constructed a surprising "limit case," a formal structure whose "geometry" or "syntax" limited the possible interpretations it could be invested with. In the case of big "T," arithmetic truth, he was able to prove in his incompleteness theorem that the logical system he had constructed could successfully capture the concept of formal proof but could not, on pain of contradiction, represent truth. Before the incompleteness theorem, it was possible to mistake proof for truth. Afterward, with Gödel's introduction of the "syntactically extreme" conditions of his formal system—the conceptual analogue of an atom smasher—no reasonable person could fail to see the distinction. In his contribution to relativity theory, Gödel, once again, constructed a limit case, this time for the relativistic geometrization of time. That is, he had demonstrated that in the mathematical construction of the Gödel universe, little "t," the variable that represents the temporal component of four-dimensional space-time, cannot bear the standard interpretation of time in the intuitive sense. Indeed, he proved that it cannot even be interpreted as "cosmic time," itself at most a simulacrum of the real thing.

Once again, he had been able to make a discovery because he had used his philosophical eye to isolate the essential properties that distinguish the intuitive from the formal concept, in this case, the properties that make intuitive time time, and was thus in a position—as those who took little "t" to be an analysis of intuitive time were not—to prove that these features were excluded by the very geometrical structure of the Gödel universe. Whereas in our world, it was possible—if you didn't look too closely—to confuse formal, relativistic time with time as ordinarily conceived, this identification became patently unacceptable in the extreme geometrical environment represented by the Gödel universe. What once was hidden was now revealed.

The similarities continue. Just as David Hilbert tried at first to avoid the consequences of the incompleteness theorem by inventing a new rule of logical inference out of whole cloth, so too the relativistic establishment, in the person of Stephen Hawking, tried to get around the embarrassing consequences introduced by the Gödel universe. If the annoying Gödel universe was consistent with the laws of general relativity, why not change the laws? Hawking thus introduced what he called the "chronology protection conjecture" (though a better name would have been the "anti-Gödel amendment"), which proposed a modification of general relativity whose primary goal was to rule out the possibility of world models like Gödel's, with their awkward chronologies permitting closed temporal loops and causal chains with no beginning. Despite having, as Russell noted in a different context, all the advantages of theft over honest toil, Hawking's chronology protection conjecture has won few adherents, its ad hoc character betraying itself.

Rarely Have So Many Understood So Little About So Much

If it is shocking that such a profound insight into the philosophical implications of the theory of relativity has had little impact on physicists, it is dismaying that Gödel's ideas have failed to catch the attention of philosophers. In this atmosphere of neglect, it is hardly surprising that the striking dissimilarity between Gödel's two great contributions to the dialectic of the formal and the intuitive has also gone unnoticed. Gödel was at once a mathematical realist and a temporal idealist. He concluded from the incompleteness of Hilbert's proof-theoretic system for arithmetic that the Platonic realm of numbers cannot be fully captured by the formal structures of logic. For Gödel, the devices of formal proof are too weak to capture all that is true in the world of numbers, not to say in mathematics as a whole. When it came to relativistic cosmology, however, he took the opposite tack. The consequence of his discoveries for Einstein's realm was not that relativity was too weak to encompass all that is true about time, but rather that relativity is just fine, whereas time in the intuitive sense is an illusion. Relativity, by Gödel's lights, does not capture the essence of intuitive time, because when it comes to time, our intuitions betray us. "As we present time to ourselves," he said, "it simply does not agree with fact. To call time subjective is just a euphemism." This, for Gödel, was the point of intersection between Kant's idealism and the temporal idealism implicit in Einstein's physics.

Having failed to notice the asymmetry between the two incompletenesses Gödel discovered, his colleagues in the relativistic and philosophical establishments were of course in no position to comprehend it. It remains one of the most important unanswered questions in our understanding of Gödel's philosophy. A promising line might proceed as follows. In the case of his incompleteness theorem, Gödel could compare the well-determined set of theorems of formal arithmetic with the equally well founded deliverances of intuitive, i.e., unformalized, arithmetic, accumulated over millennia by the world's great mathematicians, from which no contradictions have been derived. Even the concept of set, as employed "naïvely" by mathematicians, has not led to paradoxes. "This concept of set," Gödel pointed out, "according to which a set is anything obtainable from the integers (or some other well-defined objects) by iterated application of the operation of 'set of,' and not something obtained by dividing the totality of all existing things into two categories, has never led to any antinomy whatsoever." Russell's Paradox, in contrast, arose precisely from attempts like Frege's to formalize Cantor's intuitive theory of sets by "dividing the totality of all existing things into two categories," those that fall under a given concept and those that don't. "These contradictions," Gödel reminded us, "did not appear within mathematics but near its outermost boundary toward philosophy." It is formalisms like Hilbert's and Russell's that are problematic; everyday mathematical practice is not founded on a mistake.

Things stand otherwise with time. Whereas special and general relativity are coherent, well-formulated, well-understood physical theories that have enjoyed extensive empirical confirmation, our ordinary, pretheoretical, conceptions of time, i.e., of the A-series, cannot be trusted. The proof of this comes from the fact that our own experience of time in the actual world as something that lapses might well be indistinguishable from how one would perceive "time" in the Gödel universe, in which intuitive time, which lapses, is provably absent. If a form of experience is compatible with both a thesis and its antithesis, it cannot be taken as reliable testimony for either. The fact, then, that the theory of relativity fails to account for the deliverances of our everyday experience of time suggested to Gödel not that Einstein's theory is incomplete, but rather that our sense of intuitive time is founded on a misunderstanding or misapprehension. In the clash between Einstein and everyday experience, it is experience that has to yield.

Such an answer to the fundamental question of Gödel's asymmetrical responses to his two incompletenesses has not heretofore been proposed, for the very simple reason that the question itself has never been raised. The failure of his contemporaries—and ours—to appreciate what Gödel has accomplished with his Einsteinian inheritance is a sad tale indeed. Rarely have so many understood so little about so much. Gödel's "detour" into relativity has been dismissed as a bit of intellectual dabbling by someone outside his field and out of his depth. No one has seen this work for what it was: a continued development of Gödel's program of probing the limits of formal methods in capturing intuitive concepts, a move from the big "T" of mathematical truth to the little "t" of relativistic time. As a consequence of this failure, no one asked why Gödel's responses to his incompleteness results in the two cases were diametrically opposed.

The details of Gödel's conclusions about little "t" were also neglected. Cosmologists questioned whether the possibility of time travel in the nonexpanding Gödel universe was consistent with relativity, but made little note of the primary purpose for which he had constructed these world models, which was to show that since time travel was possible, time was not. And when it became clear that his new world models were indeed relativistically consistent, attention was diverted once more from the essential to the inessential: now cosmologists asked whether the actual world is an expanding Gödel universe. The foundation of Gödel's case for temporal idealism, his modal argument from the possibility of the Gödel universe to the nonexistence of time in the actual world, disappeared from sight.

Who's Kurt Gödel?

Though misunderstood and underappreciated, Gödel's birthday present for Einstein did attract some immediate attention, if not from philosophers then at least from the guardians of relativity. Of the two great theories of modern physics, general relativity is clearly the philosophical cousin, leading naturally to speculation on the origin, shape and fate of the universe—a highly theoretical, not to say metaphysical, preoccupation of philosophers from time immemorial—whereas quantum mechanics has immediate implications for technology and practice, from lasers and microchips to the whole panoply of information-theoretic hardware. Within the confines of general relativity itself, moreover, the question of time represents an especially elusive philosophical corner. What to do with time in special relativity is easy (if you know what to look for); what to do with it in general relativity is something else entirely. Since Gödel's discoveries concerned an even more isolated niche of this already remote corner—namely, speculations about geometrically extreme cosmologies with bizarre chronological consequences, not to mention Gödel's even more arcane philosophical reflections based on these monstrous models—it was to be expected that the small ripple raised by his rarefied achievements would soon fade away.

Into this quiet pond, one spring day decades later, stepped the physicist John Wheeler, a colleague of Gödel's at Princeton. He was with his friends and fellow physicists Kip Thorne and Charles Misner, with whom he was completing what would become one of the great texts in general relativity, called simply *Gravitation*. The sunshine beckoned and the three betook themselves across campus to the grassy knolls of the institute, there to meet Wheeler's friend Kurt Gödel. The warmth of the day notwithstanding, the old logician was found wrapped in his overcoat, the electric heater in his office turned on. Wheeler and friends had a question. Could Gödel shed light on the relationship between his incompleteness theorem and Heisenberg's uncertainty principle? No. For Gödel, it was bad taste even to pose such a question. Heisenberg's principle was the finest flower of the Copenhagen interpretation of quantum mechanics, itself the blue-eyed boy of positivism. It represented the high-water mark of indeterminism in physics—in effect, a rejection of Leibniz's principle of sufficient reason, so beloved of Gödel—and the acme of irrealism in physical science. As such, it was the very thorn on the rose for both Einstein and Gödel. As Gödel put it, "in physics . . . the possibility of knowledge of objectivizable states of affairs is denied, and it is asserted that we must be content to predict the results of observations. This is really the end of all theoretical science in the usual sense." The incompleteness theorem, in contrast, was a definitive refutation of positivism. Its methods and formal conclusions, though positivistically acceptable, were of a piece with classical mathematics. Moreover, the proof itself, by Gödel's lights, constituted strong evidence in favor of realism in mathematics. To have suggested a connection or correlation between Heisenberg and Gödel was a major faux pas.

Gödel notwithstanding, however, Wheeler and his friends were not far off the mark. It cannot be denied that there are striking parallels between Gödel's incompleteness and Heisenberg's uncertainty (though tact should have counseled against pointing this out to Gödel). For one thing, both thinkers were at pains to use methods that would be epistemologically acceptable to the most hard-headed positivist: formal systems in the case of Gödel's theorem, direct empirical observations in the case of Heisenberg's principle. Furthermore, each theorist drew ontological conclusions from epistemological premises, conclusions that established the intrinsic limitations of the epistemologically acceptable methods they had employed. This form of argument is the very hallmark of positivism. It is also characterizes Einstein's special theory of relativity, a fact with which Heisenberg tried (unsuccessfully) to impress Einstein. That the conclusions Gödel drew pointed to mathematical realism, while Heisenberg made the case for physical irrealism, does not alter the fact that both thinkers blazed an ontological trail through the thickets of epistemology, and that each inaugurated thereby an intellectual revolution whose full implications are yet to be realized. Not for nothing did Gödel's colleague at the institute, Freeman Dyson, remark that "the two great conceptual revolutions of twentieth century science [are] the overturning of classical physics by Heisenberg and the overturning of the foundations of mathematics by Gödel."

Now Gödel himself had a question. Would there be a discussion in their new text of the rotating universes he had discovered in relativity? No. Gödel was disappointed. He was still seeking to discover whether the actual world is a (expanding) rotating Gödel universe. The evidence for universal rotation, should it exist, would be found in the axes of rotation of the surrounding galaxies. Wheeler was taken aback by the practical astronomical preoccupations of the great logician. Gödel, he noted, "had taken down the great Hubble photographic atlas of the galaxies, lined up a ruler on each galactic image to estimate the galaxy's axis of rotation, and compiled statistics of the orientation." The results, however, were negative.

That Gödel had made discoveries about rotating universes in general relativity had been known to Wheeler for many years. He was present in 1949 when Gödel lectured on the subject at Einstein's seventieth birthday celebration. Yet he too, despite his impressive credentials, seems to have misunderstood what Gödel was saying. "In a universe with an overall rotation," he wrote, attempting to summarize Gödel's lecture, ". . . there could exist world lines (space-time histories) that closed up in loops. In such a universe, one could, in principle, live one's life over and over again." Wheeler, unfortunately, has conflated a temporal circle with a cycle, precisely missing the force of Gödel's conclusion that the possibility of closed, future-directed, timelike curves, i.e., time travel, proves that space-time is a space, not a time in the intuitive sense. Whereas a circle is a figure in space, a cycle is a journey undertaken along a circular path, one that can be repeated, in Wheeler's words, "over and over again." Exactly how many times, one wants to ask Wheeler, is the journey supposed to be repeated? The question clearly cannot be answered, since the time traveler's journey is not over time, along the closed timelike curve: it is the curve itself. Just as one cannot ask of a circle how many times the points that constitute that figure have gone around, one cannot sensibly ask how often the time traveler in the Gödel universe has made his or her trip.

Wheeler should have known better. As he himself pointed out, an "unsettling consequence of Einstein's 1905 special theory of relativity is that time is relative." And not just relative, but "static," for "the other thing that special relativity did for time is join it with space into the four-dimensional entity space-time . . . [and] a consequence of this new space-time view is that motion through time, or motion of time . . . is replaced by static time." But, as Gödel showed, a time that is relative or static is no time at all. Wheeler seems reluctant to call a spade a spade. Yet he does entitle his chapter "The End of Time," so perhaps he does, after all, recognize this. Not at all. What Wheeler means by "the end of time" is not that it disappears in Einstein's theory as a consequence of being relative and static, but rather that, as he sees it, when the "Big Crunch" comes, after the "Big Bang," time will come to an end. "There was no 'before' the Big Bang," he writes, "and there will be no 'after' after the Big Crunch." Moreover, "every black hole brings an end to time and space . . . as surely as the Big Crunch will bring an end to the universe as a whole." What Gödel has seen, it seems, Wheeler has not.

A year after he introduced Misner and Thorne to Gödel, Wheeler found himself in the office of a colleague, the cosmologist James Peebles. In walked Peebles' student Dan Hawley, announcing that he had just completed his dissertation on the question of a preferred rotation among the galaxies. Gödel, Wheeler commented, would be pleased. "Who's Gödel?" asked Hawley. "The greatest logician since Aristotle," Wheeler replied. And much more. A phone call to Gödel allowed Wheeler to apprise the greatest logician since Aristotle of the new work being done in Princeton on the rotation of the galaxies. Gödel's queries, however, were soon too demanding for the physicist, so Wheeler handed him over to the student of cosmology. The questions quickly exhausted him, too, so the phone was passed yet again, this time to Peebles. When the conversation finally concluded, there was just one thing Peebles had to say: "My, I wish we had talked to him before we started this work."

Though the world at large had not yet taken note of what Gödel had accomplished in Einstein's backyard, there were rumblings among the cosmologists that something new was brewing. Just what this was, however, would remain hidden for years to come. That a noted cosmologist was moved as recently as the 1990s to protect chronology from the Gödel universe suggests that the world is still not ready for Gödel. Yet the mere fact that as distinguished a theorist as Stephen Hawking believed protection was needed, combined with the fact that his chronology protection conjecture has so far failed to attract a significant number of adherents, suggests that readiness may be near. The zeitgeist, as Gödel noted, has its own time and agenda.

8 | Twilight of the Gods

We live in a world in which ninety-nine per cent
of all beautiful things are destroyed in the bud.
KURT GÖDEL

It all began with geometry. "Those ignorant of geometry shall not enter," Plato had inscribed over the entrance to his academy. In a passage admired by Gödel, he says in Book 6 of *The Republic* that when students of geometry "make use of the visible forms [of geometric figures] and reason about them, they are in fact thinking not of these but of the ideals [i.e., "ideas" or forms] they resemble." Thus was the path cleared for Euclid, who succeeded—not perfectly, as Kant thought, but to a considerable extent—in capturing those geometric forms in a system of axioms that remains the paradigm of theoretical knowledge, in mathematics and logic no less than in physics.

Einstein, who had the courage to employ an alternative to Euclid's system to describe the actual world, was one of the first to grasp the difference between geometry as a formal science of deduction and geometry as an empirical account of physical space, a distinction he elaborated with gusto in his essay "Geometry and Experience." He had begun to follow in the footsteps of the Greek philosophers early in life, when his youthful imagination was captured by his "holy geometry booklet." Geometry, too, was the secret password for entrance to the Gödel universe, a password Einstein himself was hesitant to invoke, yet it was also the key Einstein himself would employ to unlock the secrets to his unified field theory, a key no one but he cared to turn. "Einstein's now abandoned dream of a geometrical unification of the forces of nature" is how John Wheeler described it decades later, in 1980.

Everything Is Something Else

Turning to geometry one more time for the solution to his final problem was for Einstein a case of going home with the girl he had brought to the dance. In creating relativity theory, he had inaugurated the geometrization of physics. The mysterious limit velocity of light was to be accounted for not by ad hoc mechanical devices like the strange shrinking behavior of measuring apparatuses, but rather by the geometrical structure of space-time itself, a structure that has the limiting velocity of electromagnetic signals built into its very definition. Similarly, the force of gravity was explained not, as with Newton, as a mysterious instantaneous action at a distance that moves through an even more mysterious world-filling yet invisible substance known as ether, but rather by the geometrical device of the curvature of space-time. Time itself had been tamed—or so it seemed—by its transformation into space, into the temporal component of four-dimensional space-time. In his way, Einstein turned out to be no less an ironist than Gödel. Everything is really something else: time is really space, gravity is really geometrical curvature, energy is really mass. How can one not love such a thinker?

That Einstein and Gödel would meet on the field of geometry was altogether fitting. The parallel lines of their careers converged on the Gödel universe, at once Gödel's birthday present for his best friend and his entrance into Einstein's arena of battle. Gödel had carried Einstein's geometrization of time to a surprising conclusion, forcing us to question not just the truth but the very meaning of the Einsteinian starting point. The two walking companions had marched so far ahead of the rest of us that no one could tell whether it was they or we who were lost. Gödel's writings on Einstein did, however, provide cover for the attempt by Gödel's friends to cheer the great logician, whose health, shortly after he completed his contribution to the Schilpp volume, became imperiled by a bleeding duodenal ulcer. Promotion to full professor would not come until later, when the objections of some of his colleagues could be overcome. ("One crazy man [himself] on the faculty" is enough, said the mathematician C.L. Siegel.) A solution was found when institute director Oppenheimer, who was on the selection committee for the first Einstein Award—to be presented every three years, on the physicist's birthday—suggested that it would be fitting to divide the honor between Gödel and Julian Schwinger, the physicist from Harvard who would soon earn a Nobel Prize for his work in quantum electrodynamics. Thus on March 14, 1951, Einstein's birthday, after Von Neumann had delivered a brief speech in which he described Gödel's work as "a landmark which will remain visible far in space and time," the aging physicist was able to return the favor of Gödel's birthday gift of 1949 by personally handing his good friend the first Einstein Award. ("You deserve it," he said to Schwinger; "you don't need it," he remarked to Gödel, who needed it most of all.)

This was the first formal academic honor Gödel had ever received. In due course he would get others, including honorary degrees from Yale and Harvard (though not from Princeton; the invitation came too late) and the National Medal of Science, but he had already begun to withdraw from academic and social life. When he delivered the prestigious Gibbs Lecture that same year, 1951, before the American Mathematical Society, the only logician ever to do so, it was the last talk he would ever deliver to a mathematical audience, and one of the last he would even attend. He had never, in any case, found much profit in attending formal talks. "I never go to lectures," he said, "because I have difficulty in following them, even if I am well acquainted with the subject matter." For the remainder of his life, he would publish no new essays, not even the Gibbs Lecture, which appeared only posthumously. His sun rose and set at the same moment.

The Need for Roots

For Einstein, the darkness had long since descended. After presenting his award to Gödel, he had few years left to live. No new conquests graced his final decades. His last grand geometrical move, like Gödel's, had been followed by no one. He became preoccupied with the attempt to mitigate the very forces he had helped set into motion: the positivism, inspired by Mach, that had set the stage for special relativity; the bold decision to take Planck's quantum as a genuine aspect of reality rather than a mere calculating device; the courageous proposal, following Boltzmann, that probability be taken seriously in physics; the seminal work in forging a new quantum mechanics. Politically, he provided bookends to the sudden intrusion of physics into global politics. On one side was his recommendation to FDR that nuclear power, in the form of a bomb, be exploited to defeat his former homeland. On the other—his final act on the public stage—was his signature on a manifesto written by Bertrand Russell demanding worldwide nuclear disarmament. His white whale, however, remained his never-ending, never-succeeding search for a unified field theory, together with his attempt to find a philosophical flaw in the Copenhagen interpretation of quantum mechanics. Like Ahab, he took the hunt personally and was fully prepared to go down with the ship.

Increasingly, he withdrew from the physics establishment to pursue the great beast in isolation. After he arrived at the institute, he never visited Europe again. He never drove a car and never flew. His circle of friends diminished, with Gödel the brightest star in his shrinking firmament. Never again would he enjoy the intellectual camaraderie that had formed a cloak against all the ugliness that beset his years in Berlin. Toward the end of his life he confessed that his strongest personal ties, including those to his wife and children, had all been failures. When his wife Elsa's daughter from a previous marriage, Ilse—whom he had once thought of marrying—lay dying of cancer in Paris in 1934 at the tender age of thirty-seven, he declined to accompany his wife to attend to her. His first wife, Mileva Maric-

Einstein, died alone in Zurich, desperately unhappy, unreconciled with the man who had left her. His daughter with her, Lieserl, born out of wedlock, disappeared into the mists of time. His gifted first son, Eduard, became schizophrenic and was deposited in a psychiatric clinic, where he remained for the rest of his life, unvisited by his father. His second son, Hans Albert, always distant, remained so after he too emigrated to America. And the second marriage, finally, of Einstein, like the first, was no success, though it did at least provide a slender root in an otherwise rootless existence. Its removal in 1936, yet another deracination, deeply affected him, surprising Elsa. "I never thought he loved me so much," she told her friend Antonina Vallentin, "and that comforts me." Sad words, indeed, from a dying spouse. With Elsa's death his personal universe collapsed in on itself.

For the remaining years of his life his most visceral human connection was to what he called his "tribe," his fellow Jews, the deepest root of this rootless man (though he somehow never managed to plant himself in the land of Zion). "My relationship to the Jewish people," he wrote, "has become my strongest human bond." Why he chose this for his fundamental human tie remains to be explained. One suspects that the French philosopher Simone Weil's dark study, *The Need for Roots*, contains greater hints than are found in the standard literature, which has difficulty acknowledging the degree of discord between Einstein's self-confessed "tribalism" and his lifelong commitment to rationality and internationalism.

When the end finally came at the age of seventy-six, Einstein could not help feeling embarrassed at the larger-than-life icon he had become. In March 1955, a month before he took his final voyage, he confided to his long-time friend, Queen Elisabeth of Belgium, that "the exaggerated esteem in which my lifework is held makes me very ill at ease. I feel compelled to think of myself as an involuntary swindler." The day before he died, he requested his latest version of unification theory and proceeded to make some calculations. He did not put up a fight to remain living. "It is tasteless to prolong life artificially," he told Helen Dukas; "I have done my share, it is time to go."

The Absence of the Muses

According to the mathematician Stanislaw Ulam, Gödel too, toward the end of his life, nurtured fears that his contribution had been overestimated. Gödel, Ulam said, had "a gnawing uncertainty that maybe all he had discovered was another paradox à la Burali-Forte or Russell." Gödel himself, however, denied this. "Ulam wrote a book . . . [in which he] says that perhaps I was never sure whether I had merely detected another paradox like Burali-Forte's. This is absolutely false. Ulam doesn't understand my result." Since Ulam is not here to defend himself, we cannot determine whether it is his memory or Gödel's that is at fault. Others confirm, however, that Gödel, like other distinguished thinkers who joined the institute, occasionally wondered whether he had done enough to justify his appointment.

Like Einstein, Gödel had led a life of increasing isolation and reclusiveness since coming to the institute, a tendency that only increased after he received the Einstein Award and delivered the Gibbs Lecture. He too spent his final years in a lost cause, part formal and part philosophical, searching for new axioms to decide the continuum hypothesis (and thus settle the question of whether there is an infinity between the number of points on a line and the cardinality of the natural numbers), and seeking a definitive refutation of the thesis—bolstered by Cohen's independence result for the continuum hypothesis—that the results of mathematics are in some sense only the reflection of human convention. This was a theme he pursued in his Gibbs Lecture—in which he invoked his own incompleteness theorem as evidence for his Platonism—as well as in his contribution to yet another Schilpp volume, devoted this time to Rudolf Carnap, his old friend and foe from the Vienna Circle. But he was never satisfied with this essay and did not allow it to be published.

Gödel and Einstein, two great thinkers each of whose earlier years had been marked by a string of successes that left their contemporaries breathless, spent their final decades in a doomed commitment to lost causes. What happened? Why were the muses absent during the twilight of the gods?

No one, of course, has ever divined the secret of the muse (else we would all become Mozart), but we can nevertheless note certain salient factors in the striking lack of success Einstein and Gödel enjoyed in their final years. What does it take to make a great scientific discovery? Two elements are crucial. One must have an insight into which problems are ripe for resolution, and one must then have the craft—or invent it—to solve the problem one has had the audacity to recognize as solvable. Both elements, clearly, were present in Einstein's success with relativity theory and his early work in quantum mechanics. With regard to the first element, his biographer has pointed out that "in 1905, work on spectral lines could not have gone beyond an attempt at a phenomenological interpretation, even for Einstein. The fact that he did not attempt it show[ed] him to be a master of the art of the soluble." Both were prominent in his discovery of general relativity. No less a figure than Planck himself warned Einstein that it was hubris to attempt to rethink gravity after three hundred years of Newton. And the technique required to forge the new theory turned out, unlike special relativity, to require highly nontrivial mathematics that strained Einstein's formal capacities almost to the breaking point.

For Gödel, too, each element had been present. In the incompleteness theorem, he understood that it was possible to test the limitations of formal systems, undermining the confidence in purely deductive methods, inaugurated by Euclid, that had held sway for more than two thousand years. Finding, or rather creating, the methods needed to resolve this now solvable problem—including Gödel numbering and the arithmetization of metamathematics—was perhaps Gödel's chief boast. The continuum hypothesis, too, which had defeated its inventor, Cantor, yielded (at least in part) to Gödel's realization that its consistency with the axioms of set theory could now be settled, if one had the ingenuity to employ the new set-theoretical techniques Gödel had managed to cook up.

For other distinguished scientists, too, the same pattern for scientific discovery holds. The secret to James Watson and Francis Crick's discovery of the structure of DNA (leaving aside the small matter of Rosalind Franklin's desk drawer) was as much their realization—almost unique at the time—that the problem was now solvable, as it was their technical competence in fitting together all the pieces that lay scattered about, unconnected, in various workshops. In their case, the first step was probably the crucial one. They were not the only scientists equipped to solve the riddle once it was understood to be solvable; the great Linus Pauling was pursuing the problem with equal fervor but, unfortunately for him, the wrong idea. This partly explains Watson and Crick's frantic anxiety to find a solution at breakneck speed and their obliviousness to the niceties of professional ethics.

More recently still, Andrew Wiles's dramatic solution to the problem of Fermat's last theorem also conforms to the pattern. Wiles himself has written of the moment he realized that the theorem could now be proved, that we could really get there from here, and that apart from the usual cranks, he was alone in knowing this. Secreting himself away in his attic for years, he was able finally to bring forth the second element, the virtuoso technical methods with which all the pieces of the puzzle uncovered by his great predecessors could finally be sewn together. (Even Wiles, however, it turned out, could not put all the pieces together without help, after his initial proof turned out to contain a flaw.)

In the case of Gödel and Einstein, there is no indication that their ability to cook up the second half of the recipe for scientific success had dwindled. Both men remained mathematically nimble to the very end. It was with the first element that problems arose. They had simply bitten off more than they, or anyone else, could chew. To this day, half a century after Einstein's failed efforts, we still do not have a clear path to the unification of the very small with the very large, quantum mechanics with relativity. If, as some suspect, the most promising avenue lies in string theory, with its exotic mathematics of ten dimensions, then Einstein clearly never had a chance. There is simply no way even Einstein in his day could have dreamed up string theory. As for quantum mechanics, while Einstein's philosophical objections retain their power to haunt physicists—"I cannot define the real problem," said Feynman in 1982, about the EPR paradox, "therefore I suspect there's no real problem, but I'm not sure there's no real problem"—the final philosophical account of the nature of quantum reality (or unreality) has yet to be written. Not only are we not there yet; no one seems to know where we're going or how we will know when we get there.

Gödel's lost cause—finding new axioms that will settle, in a convincing, non ad hoc, manner, the continuum hypothesis—has likewise seen little progress, either to suggest wherein the answer lies or even to indicate whether there will ever be a definitive answer. Gödel had finally tackled a problem that was anything but ripe for resolution. Nor has there been any breakthrough on settling the companion philosophical problem of the extent to which mathematics represents a reality independent of human convention. The appearance, posthumously and in different versions, of Gödel's contribution to the Schilpp volume on Carnap, entitled "Is Mathematics Syntax of Language?", has revealed the intricacies of Gödel's attempts to settle accounts with his old friend from Vienna, but it has not adduced arguments that command universal assent. The essay represents a struggle more than a consummation.

Sowing without Reaping

Gödel had received the invitation to contribute to the Carnap volume in 1953, just two years after he delivered his Gibbs Lecture, "Some Basic Theorems on the Foundations of Mathematics, and Their Implications." The common theme was to find a convincing argument in favor of Platonism and against conventionalism or formalism, using his incompleteness theorem as a powerful new weapon in the war. But Gödel soon found that not even his superweapon could blast a shortcut through the tangled thickets of mathematical ontology, an insight Wittgenstein—much as his aims diverged from Gödel's—had reached on his own years earlier, in his attempt to remove the spell that Gödel's theorem had cast over philosophers. One might caricature Wittgenstein's conclusion in this way: whereas (parts of) mathematics possess a beguiling symmetry, philosophy will always be to some extent messy and ugly. In his youth, he had gushed over the beauty of Russell's *Principia Mathematica*, comparing it to music. In his later years, he took the other side, drawing attention to what he called the "motley" of mathematics. Gödel, in contrast, kept faith that the beauty of mathematics could be matched by philosophy. His contribution to the philosophy of time gave support for this belief, but he was well aware that his drafts of the Carnap paper, with their never-ending revisions, were anything but pretty. None, in his estimation, was worthy of publication. "In view of widely held prejudices," he finally wrote to Schilpp in 1959, "it may do more harm than good to publish half done work."

The ontological project *contra* Carnap was never finished. Nor was his attempt to complete the other half of the philosophical coin, epistemology. Unlike other mathematical Platonists such as Frege and Church, Gödel understood the need to supply his mathematical ontology with a convincing epistemology. Here he turned for help to Frege's contemporary and fellow philosopher, Edmund Husserl. He became a profound (if unhappy) student of Husserl's recalcitrant texts, a project that consumed ever-greater amounts of his time and energy. He counseled his surprised colleagues to do likewise, with what success, one can imagine. But here too, by life's end, though he had made considerable progress, nothing definitive emerged.

Nor, for the remainder of his life, was Gödel able to convince the physics or philosophy community that he had achieved a breakthrough on relativity theory's philosophical consequences regarding the existence of time. His contribution to the Schilpp volume on Einstein may have provided cover for him to be offered the Einstein Award, but it had signally failed to establish his bona fides as a philosopher, and it did nothing to turn philosophers' attention to the burning question of the reality of time.

Suspicions of Piety

Gödel's attempt to discover the truth about the abstract universe of sets and numbers had stalled, as had his assault on the starry heavens. Undeterred, he aimed still higher. He tried to pin down God Himself, developing his own version of the Anselm-Descartes-Leibniz "ontological argument" for the existence of God, a being, by hypothesis, so perfect, if His existence is possible at all, He must exist not just in the actual but in every possible world. The step from God's possibility to His actuality was relatively straightforward, given a suitable choice of axioms for one's "modal logic" (i.e., the logic of the modes of possibility and actuality). The hard part, Gödel realized—as had his hero Leibniz before him—was proving that a divine being was so much as possible. This Gödel attempted to do via a highly compressed formal argument, which, once again, he declined to publish. He feared, he told his friends, that its publication might suggest to his skeptical philosophical colleagues that he actually believed in God, whereas (he claimed) in fact it was a mere formal exercise.

His assessment of the religious inclinations of the philosophical community was probably accurate: "Ninety per cent of contemporary philosophers," he wrote to his mother in 1961, "see their principal task to be that of beating religion out of men's heads." Charles Parsons, a philosopher and logician at Harvard, tells a story that speaks to Gödel's concerns. During an interview in 1955 for membership in the prestigious Society of Fellows at Harvard, where he was a first-year graduate student, he let it slip that he had done readings in theology and found Pascal interesting. Even though, he insists, "I was not then and never have been a Christian," he had forgotten that one of his examiners, the dean of American philosophers, W.V.O. Quine, "was a firm opponent of religion." Legend has it that when the meeting concluded, Quine was heard to say, "Good grief, Parsons is pious." Needless to say, he was not elected. (He was, however, invited to reapply.)

Gödel was no more successful in preventing himself from being considered pious. Word leaked out about his proof, and no one, then or now, was fooled into thinking it was a mere "formal exercise." When the proof itself finally materialized, posthumously, problems were found in the details. Whether repair is possible is an open question, as is the problem of whether an amended proof, with its revised premises, would be convincing. What is beyond dispute, however, is that the appearance of Gödel's version of the ontological argument has had little effect on the confidence of philosophers that a formal demonstration of God's existence is impossible. Knocking religion out of people's heads continues to be a favorite philosophical pastime.

Preoccupied with not appearing eccentric or out of fashion to his colleagues, Gödel was nevertheless universally perceived as both. What the logician Solomon Feferman has characterized as Gödel's "special caution" had one especially unfortunate effect: it kept him from contributing to important branches of logic and mathematics that he himself had been instrumental in inaugurating. As Feferman points out, Gödel—perhaps exaggerating the continued influence of the Hilbert school and the dominance of positivism—having demonstrated the indefinability of arithmetic truth within formal arithmetic, declined to go on to provide a formal account of the concept of truth itself. That notable task was left to his colleague Alfred Tarski, with whose name the subject is now identified. Similarly, having laid the groundwork for much of the theory of effective computability in his seminal discussions of recursive functions in his incompleteness theorem, Gödel declined to provide a definitive account of effective computability, a central concept of today's theoretical computer science. That task was accomplished by Alan Turing, with contributions from Church, Stephen S. Kleene and J. Barkley Rosser. "One may wonder," says Feferman, "how logic might have been different had Gödel been bolder in bringing his philosophical views into play in relation to his logical work."

"The World Tends to Deteriorate"

Gödel's paranoia, then, undoubtedly deprived the world of his contributions to important areas of modern thought. In time it would deprive him of life itself. Though he recuperated well enough from the bleeding ulcer he suffered in 1951, cheered by the Einstein Award and by the invitation to deliver the Gibbs Lecture, his mental and physical state would soon enter a downward journey from which he would never recover. In the coming decade the colleagues who had been closest to him, who had shepherded him since he first arrived in Princeton, would all die: Einstein in 1955, Von Neumann in 1957, Veblen in 1960. Einstein had kept it secret that he had a life-threatening heart condition—an aneurysm, a weakening in the abdominal aorta, diagnosed in 1948—that could take him at any moment, and when that moment finally arrived, Gödel was stunned. His thoughts, always gloomy, took on a darker hue. "We live in a world," he wrote, "in which ninety-nine per cent of all beautiful things are destroyed in the bud."

He had studied Hegel and developed his own philosophy of history, according to which the world is subject to large-scale "noncausal" laws: "There are structural laws in the world which can't be explained causally." These did not, in his view, justify the post-Enlightenment, Christian belief in human progress. "The world tends to deteriorate," he wrote. "Good things appear from time to time in single persons and events . . . but the general development tends to be negative." Christianity, with which he was generally sympathetic, was no exception. It "was best at the beginning. Saints slow down the downward movement." As Simone Weil put it, although "since [Christ's] day there have been no very noticeable changes in men's behavior," "drops of purity" appear from time to time. Philosophy, according to Gödel, suffered a similar fate: "Philosophy tends to go down." Indeed, "it is, at best, at the point where Babylonian mathematics was."

Unsurprisingly, Gödel had few interactions with the Babylonian philosophical establishment in Princeton or elsewhere. With most of his closest friends dead and Adele suffering from increasingly debilitating ailments (variously characterized as hypertension, arthritis, bursitis and gall bladder disease), he turned in his final decade to his old acquaintance Oskar Morgenstern, who was granted the dubious privilege of witnessing Gödel's gradual descent into full-blown paranoia and hypochondria. Where once Adele had been there to tend to his needs, it was left henceforth to Gödel to care for his wife.

By 1968, Morgenstern had grown alarmed at how gaunt his friend had become. It became nearly impossible to persuade him to eat, with predictable consequences. Hypochondria joined forces with paranoia, and soon Gödel was claiming that his doctors were lying to him, their medications misidentified, their textbooks ill-written. His distrust of doctors was combined with an all too realistic fear that they would commit him to a psychiatric hospital. Soon he failed even to recognize them, and Morgenstern in turn could barely recognize his friend, who was now hallucinating and looked like a "living corpse."

Princeton too, according to Gödel, was against him, and Morgenstern, who had thrice failed to get the university to award his friend an honorary degree, had a hard time convincing him otherwise. When he tried to calm Gödel by assuring him that at least he was a true friend, Gödel replied, sadly, that a real friend would have brought him cyanide. (Alan Turing, it should be recalled, succeeded in doing away with himself by this means, using a syringe to squirt the poison into an apple, imitating Gödel's favorite fairy tale.) The hallucinations continued, as did the appeal for assistance in committing suicide.

The imaginary health problems were made worse by real ones. In 1974, Gödel's enlarged prostate blocked his urinary tract, a serious problem exacerbated by his refusal to seek treatment. Only when the pain became unbearable did he finally check into a hospital, where he was catheterized. But although surgery was recommended, he refused, rejecting the diagnosis. He removed the catheter himself, which had to be forcibly reinserted. He never did relinquish his opposition to surgery, and in the end remained permanently catheterized, leaving him in a condition of constant discomfort that could only intensify his depression. Into this fragile, pain-suffused body, Gödel proceeded to insert a whole cabinet of medications, which Morgenstern was amazed his friend could survive. Gödel ingested milk of magnesia, Metamucil, Keflex, Mandelamine, Macrodantin, Gantanol, Achromycin, Terramycin, Lanoxin, Quinidine, Imbricol, and Pericolase. The only thing he was reluctant to admit into his starved body was food.

Despite bouts of hallucinations and a constant fear of people, he had moments of great lucidity and even charm, and managed a few extended human contacts, the most important being his association with the logician Hao Wang. Fortunately for posterity, Wang took it on himself to coax out of Gödel his unwritten philosophy. From 1971 to 1972, the two met at Gödel's office every other Wednesday for two hours. If they could not meet in person, the discussion was carried out by phone. The accounts of these exchanges and Wang's observations on them, published after Gödel's death, are valuable documents, though it is unclear to what extent Gödel was trying out ideas, or musings, sufficiently far-fetched that he would have been reluctant to own up to them in print. What the philosopher Richard Rorty has aptly said of Plato also applies to Gödel: we are still struggling to separate out the straight lines from the jokes.

When none of his remaining friends could persuade Gödel to eat, his demise was guaranteed. In late December 1977, weighing sixty-five pounds, he was finally admitted to Princeton Hospital. He died on January 11, 1978, from what was diagnosed as "malnutrition and inanition" due to "personality disturbance." They buried him in Princeton Cemetery. Adele, whose capacity to save her husband from himself had long since been exhausted, lived on until 1981.

A friend for life, Einstein did not lie beside him in death. Fearing that his grave would become "a place of pilgrimage, where pilgrims would come to view the bones of a saint," he had asked to be cremated, his ashes scattered. Gödel had no such fears. There was a memorial service at the institute, where Wang paid tribute to his colleague, as did the mathematician André Weil, brother of Simone Weil, whose dark assessment of human history so strongly resembled Gödel's. His passing, unlike Einstein's, attracted little attention. The disappearance of this giant planet disturbed no other orbit. To philosophers, then as now, he was a simply a logician trying to pass as a philosopher.

9 | In What Sense Is Gödel (or Anyone Else) a Philosopher?

Engaging in philosophy is salutary, even when no positive results emerge. . . . The color is brighter, that is, reality appears more clearly as such.

KURT GÖDEL

It is difficult to protect our interests while we are alive. Much more so when we're dead. Gödel, throughout his academic life, was exceptionally anxious to avoid being considered a philosophical dilettante or crank, and a "pious" one at that. In this, he failed completely. His "special caution" (as Feferman described it) succeeded only in keeping him from contributing to important fields of research. His refusal to publish his ontological argument for the existence of God fooled no one. He could not hide the fact that he was a kind of believer and that his argument, like Leibniz's before him, was anything but a mere intellectual exercise. What Russell called Gödel's "unadulterated Platonism" marked him for some as an intellectual throwback to "precritical" times, before Kant launched his "critique" of pure reason, a police operation by which Kant intended to curb the pretensions to theoretical knowledge of most, if not all, philosophers who preceded him. Even Gödel's writings on Einstein succeeded only in convincing cosmologists that strange things happen when a logician pays too much attention to the equations of relativity, while forgetting their physical meaning. His absence, in turn, from the Wittgenstein revolution—which turned philosophy toward the human construction of language and the "games" people play with it—combined with his refusal to pay homage to the preeminent figure in contemporary analytical philosophy, W.V.O. Quine, marked him as a philosophical castaway.

After Gödel's death, his colleague and amanuensis Hao Wang published excerpts of their philosophical discussions, in an attempt to give a more rounded and positive cast to Gödel's image as a thinker. He too failed. Too many philosophers had already argued to the contrary. The full extent of the damage became clear during a symposium held nearly two decades after Gödel's death to celebrate his contributions to philosophy. The celebration resembled a wake.

Who Buried Kurt Gödel?

On a cold February day in 1995, a distinguished group of philosophers, mathematicians and logicians assembled to honor Gödel in a symposium hosted by Boston University entitled "Gödel's General Philosophical Significance." The speakers included those who had been assigned the task of guarding Gödel's legacy by gathering his published and unpublished essays, with editorial introductions, in a definitive edition of the *Collected Works*. Plato's question from *The Republic*, however, hung in the air: "Who will guard the guardians?" It was clear from the start that the speakers had come not to praise Gödel but to bury him. John Dawson, however, one of the guardians, was different. He alluded to the neglect of Gödel as a philosopher of time, but drew attention to the fact that there were exceptions. As he put it in his soon to be published biography of Gödel, "To date, only a single volume (Yourgrau 1991) [*The Disappearance of Time*] has examined in any detail . . . the ramifications of [Gödel's] cosmological work for the philosophy of time." Yet neither he nor any other speaker attempted to remedy the lack of attention that had been paid to Gödel's writings devoted to the subject.

Some recounted personal anecdotes about their encounters with Gödel, while others spoke briefly about his Platonism. Two speakers, however, stood out. Warren Goldfarb, from Harvard, another guardian, addressed the question of what Gödel had succeeded in accomplishing in philosophy. The answer, according to Goldfarb, was nothing. The audience, which included Hao Wang as well as the author of the only book devoted to the ramifications of Gödel's cosmological work for the philosophy of time (myself), was stunned. A hand was raised during the question period. "Do I understand you correctly, Professor Goldfarb, that in your judgment Gödel, though a great logician, was a philosophical fool?" A polite smile was Goldfarb's only answer.

But another scheduled speaker, Burton Dreben, also from Harvard, could not restrain himself: "Wait until you hear my talk!" And in due course, Dreben delivered. He made explicit in his presentation (entitled, simply, "Gödel") what had been implicit in his colleague's talk, that Gödel was a logician trying to pass as a philosopher. For Dreben, it seemed, this was a kind of scandal. He was moved to deliver a sermon on the harm that is done when people who are good at purely formal thinking get the idea that they are also qualified to contribute to philosophy. That Dreben's own position in the philosophical world owed much to his reputation in formal logic was an irony that seemed lost on him. Being good at one task, he stressed, says nothing of your ability to succeed at the other. Fascists, he said, were sometimes good at science, but that doesn't mean we should take them seriously when they try their hand at philosophy.

Gödel, by Dreben's lights, was a throwback to a benighted, pre-Kantian era of philosophy, a vagabond in the modern vineyards painstakingly planted by the likes of Kant, Wittgenstein and Quine. To Dreben, these thinkers represented the future. Gödel was the past. What especially irked him was that Gödel had the audacity "in this day and age" to engage in rational theology, a reference, no doubt, to Gödel's recasting of the ontological argument. He seemed almost driven to despair by what Gödel had done. If Goldfarb had quietly begun the delicate task of burial, Dreben was in haste to get the body

into the ground. But would the ground receive the man being cele-
brated? By what right had Gödel, the logician, ventured to teach
philosophers about their own discipline? What kind of discipline is
philosophy, anyway?

The Philosopher on the Train

If you meet a philosopher on a train and ask him his profession, he is
likely to lie. It is not that philosophers are especially prone to lying,
but rather that philosophy is a peculiar profession. To tell your fellow
passenger that you are a philosopher opens up an awkward line of
questioning. To begin with, describing yourself as a philosopher is like
calling yourself a sage, a seeker of wisdom. We all seek wisdom, after
all, but that won't feed the bulldog. A safer response is to account
oneself a philosophy professor. This is fine, unless you happen to be an
actual philosopher, in which case it is just another lie. As the philosopher Leo Strauss once said, you are as likely to find a real philosopher
in a philosophy department as you are to discover a Picasso in the department of fine arts. (Wittgenstein, though he taught for years at
Cambridge University, is correctly described as a philosopher, not a
professor.) If you take the plunge, however, and accept the label of
philosopher, you must be prepared for the disappointment when your
listener hears that you don't live in a hut on a mountaintop, haven't
uncovered the secret of life, and cannot explain why the world exists.
If you are foolish enough to go further and attempt to describe your
lifelong attempt to reconcile the epistemology of mathematics with its
ontology, be prepared to encounter a look in which boredom and hor-
ror are blended equally. Best, therefore, to say simply that you are an
architect, and leave it at that.

Gödel, one can be certain, was well aware of how the world re-
gards those who dare to call themselves philosophers; he would have
been reluctant to describe himself that way to his fellow traveler. Being
officially a logician gave him adequate cover. Though Dreben and a

host of others took him to be a logician trying to pass as a philosopher, he is more accurately described as a philosopher posing as a logician. More than most academic philosophers, he engaged in philosophy in a manner of which Parmenides and Plato would have been proud: asking fundamental questions about the nature of time, being, death, God and the world of transcendent forms, or "ideas."

He shared, too, Socrates' and Wittgenstein's mistrust of philosophy as just another paid profession. "To do philosophy is a special vocation," he wrote. "We do see the truth, yet error would reign." The special vocation carried with it duties and dangers. "Philosophy," he said, "is a persecuted science." He was not thinking of the danger of not acquiring tenure, but rather of the risk you take if you question the ruling paradigm. Whereas "moral relativity," for example, is a widespread catchphrase that encourages us to attribute the lack of progress in ethics to the fact that philosophers have a constitutional predilection to objectify what is merely subjective, Gödel made a different, shrewder assessment: "Actually, it would be easy to get a strict ethics—at least no harder than other basic scientific problems. Only the result would be unpleasant, and one does not want to see it and avoids facing it—to some extent even consciously."

The more pressing question is whether Gödel the philosopher deserved the respect accorded full-fledged members of the academy. The answer, one feels, ought to be clear, yet academic philosophy is a most peculiar discipline. Despite tracing its lineage to such thinkers as Parmenides, Heraclitus, Socrates and Plato, philosophy's right to exist is constantly called into question, not least by its leading practitioners, from Kant to Carnap, from Wittgenstein to Quine. For philosophy as such pretends to a kind of knowledge greater than anything mere mortals seem equipped to discover. Kant put it thus in a beautiful passage in the *Critique of Pure Reason*:

> The light dove, cleaving the air in her free flight, and feeling its resistance, might imagine that its flight would be still easier in empty space. It was thus that Plato left the world of the senses, as

setting too narrow limits to the understanding, and ventured out beyond it on the wings of the Ideas, in the empty space of the pure understanding.

Sweeping aside traditional metaphysics from Plato to Leibniz as insufficiently grounded in the bedrock of sense experience, Kant offered a comprehensive account of what man can and cannot know. His critique changed the course of modern philosophy, but unfortunately it was never clear whether his account of the nature of things violated his own precepts. Can sensory experience, for example, on its own teach us what sensory experience can and cannot accomplish?

This problem seems endemic to the enterprise of constructing a philosophical system. All too often, a philosopher finds himself painted into an epistemic corner of his own making. The famous motto of the Vienna Circle, for example, that the meaning of a proposition is its method of verification, was demolished by one of their own, an astute philosopher who had once belonged to the circle. In a classic essay, Carl Hempel pointed out that since the "verifiability [or empiricist] criterion of meaning" cannot itself be verified by experience, by its own account it lacks cognitive significance. In itself, it is neither true nor false. The circle's criterion for meaning turned out to be meaningless.

A still more dramatic example comes from the first (and only) book Wittgenstein published during his lifetime, the *Tractatus Logico-Philosophicus*, in which, like Kant before him, he attempted to set limits to what can be known, or rather, to what can be said, thus helping to inaugurate the famous "linguistic turn" in philosophy. For Wittgenstein, nothing that was of genuine value—such as the beautiful, the good or the meaning of life—could actually be stated (as opposed to "shown"), and everything that could be said, which amounted to the substance of physical science, was absent of value. Displaying a greater sense of irony than had Kant, he acknowledged that the book itself attempted to say what could not be said. Strictly speaking, he admitted (or boasted), it was all nonsense. But deep nonsense. One must, he added poetically, "throw away the ladder after [one] has climbed up it." F.P. Ramsey, a brilliant young mathematician who wrote an astute review of the *Tractatus* at the tender age of twenty, was unimpressed. Though he would go on to become a close associate of Wittgenstein's, he never forgot or forgave the nonsense of the *Tractatus*. "Philosophy must be of some use," he wrote, "and we must take it seriously." If, however, he added, "the chief proposition of philosophy is that philosophy is nonsense," then we must take this seriously "and not pretend, as Wittgenstein does, that it is important nonsense." More succinctly: "What we can't say we can't say, and we can't whistle it either."

Wittgenstein would in time renounce much of the *Tractatus*, but he never recanted his skepticism about the existence of philosophy as a kind of "superscience," first among equals. The temptation to engage in deep philosophical pronouncements, he said, amounted to a kind of psychological disorder or mental cramp. The job—indeed the duty—of a genuine philosopher was to enlighten the patient by showing him that the illusion of depth was the result merely of his skating on the thin ice of confused linguistic practice. The tantalizing figures that appeared to linger deep within the ice were only reflections of the subject looking at himself. Upon awakening from his philosophical slumbers, the patient, like one of Freud's subjects, would arise from the couch, shake off his dream, and return to the sober dreariness of everyday life. Language would, in his memorable phrase, no longer be on holiday. It was hardly surprising, then, that when Wittgenstein's fellow Austrian Karl Popper addressed the Cambridge Metaphysical Society on the topic, "Are There Philosophical Problems?", his positive response roused considerable ire. Sparks flew, a poker was grabbed by Wittgenstein from a smoldering fireplace, and material was provided for a lively book eventually to be written about the incident.

Skepticism about the existence of philosophy, however, did not keep Wittgenstein away from the university. He fled once, only to return and take up residence at his old haunt, Trinity College at Cambridge University, and become for decades the dominant figure in the academic philosophical world. Unlike Socrates, Plato, Frege and Russell, but like Marx and Freud (and Jesus), he surrounded himself with a group of disciples who spread the gospel far and wide. It is an irony that precisely those philosophers who have called into question the existence of philosophy in the traditional sense have been, and continue to be, the most influential in the halls of academe. Kant the protopositivist, Carnap the actual positivist, Wittgenstein the patron saint of positivism, Quine the preeminent opponent of "first philosophy," these are the movers and shakers of modern academic philosophy in the Anglo-American world. Gödel, a disciple of none, had sealed his fate. How could he hope to be taken seriously by an academy whose founding fathers he refused to embrace?

"Precritical"

Goldfarb and Dreben, in contrast, did embrace them. And they were far from alone. "After the devastating attacks by Wittgenstein and Quine," wrote the philosopher Paul Horwich in 1990, "it is now widely believed that the sciences exhaust what can be known and that the promise of metaphysics was an intellectually dangerous illusion." At the Boston University conference, when Goldfarb rose to deliver his assessment, he did so as a colleague of Quine and a follower of Wittgenstein who clearly knew his duty regarding this "intellectually dangerous illusion." Though Goldfarb is a logician of distinction and an important commentator on Wittgenstein, the soundness of his philosophical judgment seems inversely proportional to his proximity to Gödel. Here too he is not alone.

Since Gödel is not here to defend himself, the task falls to others, but to assess the text of Goldfarb's presentation, entitled "On Gödel's General Philosophical Outlook," is difficult, since although some speakers submitted their talks for publication, Goldfarb and Dreben did not. Fortunately, however, a transcript does exist of a closely related presentation, entitled "Gödel's Philosophy," which he gave in July 1990 before a meeting of the Association for Symbolic Logic in Helsinki. Opening this talk with Gödel's view that metaphysics should ideally be presented by a small set of primitive axioms, Goldfarb commented that for Gödel, "all the important content of the primitive concepts can by exhibited in precise axiomatic relations to other concepts." Though this description is not as clear as it should be, it puts one in mind of the doctrine of "implicit definition" adopted by Hilbert, for whom the meanings of the primitive concepts in an axiomatic system are exhausted by their relationships to the other concepts. In geometry, for example, there is no more to being a line than its relationship to points and planes. Gödel, however, explicitly rejected this doctrine, as did his predecessor Frege. That is why he insisted that to know the primitive concepts, one must not only understand their relationships to the other primitives but must grasp them on their own, by a kind of "intuition."

Goldfarb continued: "There is no hint" in what Gödel wrote, he said, "that the truths of metaphysics are problematic in any special way, or pose special problems of our access to them." Thus Gödel's view, according to Goldfarb, is "precritical, in Kant's sense of 'critical.'" In fact, however, Gödel rejected Kant's critical assessment of the possibilities for systematic metaphysics, a rejection founded not on ignorance but rather on a deep understanding of Kant. Yet Goldfarb chose to describe Gödel not as "post-Kantian" but "precritical," i.e., as a philosophical naïf, not up to speed on Kant, rather than as someone steeped in Kant who nevertheless rejected much (though not all) of Kant's "critique." Indeed, the longer, original drafts of Gödel's contribution to the Schilpp volume on Einstein, entitled "Some Observations about the Relationship of Theory of Relativity to Kantian Philosophy," leave no room for doubt that Gödel had a profound understanding of Kant, which enabled him to demonstrate a striking and previously unsuspected connection between Kantian idealism and Einsteinian relativity. This newly published essay makes clear that Gödel, though he accepted certain elements of Kant's philosophy, systematically rejected its main thrust, which assimilated knowledge to the knower, not the known, and thus gave Kant's philosophy a subjectivist cast. To characterize the author of this essay as "precritical" is perverse.

Also off the mark was Goldfarb's assessment that Gödel, in his naïveté, failed somehow to appreciate the difficulty of finding the right concepts and axioms for metaphysics. Nothing could be further from the truth. Time, for example, in relation to being, Gödel considered one of the basic concepts, but he believed that the attempt to discover what is fundamental about our thinking about time can receive no assistance from physics, which, he argued, combines concepts without analyzing them. Instead, we must reconstruct the original nature of our thinking, a project fraught with difficulty. For assistance he turned not to Einstein but to Husserl and phenomenology, but he acknowledged that "the problem of time is important and difficult. For twenty-five years Husserl worked on just this one problem: the concept of time." The situation in mathematics was no different. "The way . . . we form mathematical objects," he said, "from what is given—the question of *constitution*—requires a phenomenological analysis. But the constitution of time and of mathematical objects is difficult."

Since the fundamental concepts are primitive and their meaning is not exhausted by their relationships to other concepts, how can we manage to gain some insight into them? What is required, said Gödel, is "a clarification of meaning that does not consist in defining." His fellow Platonists, Plato and Frege, had little to say on how to accomplish this. Husserl, however, who early in his career had debated Frege on the foundations of arithmetic, devoted himself to just this task, and Gödel, in turn, devoted himself to the task of discovering what Husserl had found. Husserl called his new way "phenomenology," which Gödel described as a method by which we can "focus more sharply on the concepts concerned by directing our attention in a certain way, namely, onto our own acts in the use of these concepts." If we are successful, said Gödel, we achieve "a new state of consciousness in which we describe in detail the basic concepts we use in our thought."

Exactly what this new method came down to is not easy to fathom. Gödel struggled for years to pierce the veil of Husserl's rebarbative prose, to follow him on his long, winding way. "I don't particularly like Husserl's way," he told Wang, "long and difficult." Indeed, "I love everything brief," he wrote to his mother, "and find that in general the longer a work is, the less there is in it." Yet Goldfarb appeared to have no difficulty understanding Husserl, and no qualms about dismissing out of hand what Gödel hoped to find by studying his writings. According to Goldfarb, Gödel gives phenomenology "a highly subjectivist cast" and "provides no evidence that observation of one's stream of consciousness" will give insight into the concepts one is employing. The "passing show," Goldfarb assured his audience, will never assist us in grasping concepts or finding new axioms. The great Frege had demonstrated this. "The muddled results of bare phenomenological examination," said Goldfarb, "were pointedly and effectively criticized by Frege in his review of Husserl's *Philosophie der Arithmetik*."

Neither Husserl nor Gödel, however, thought of phenomenology as merely taking note of one's stream of consciousness or gazing at the "passing show." That is the method of an empiricist like Hume, not a rationalist like Husserl. Gödel saw phenomenology as an attempt to reconstruct our original use of basic ideas, to focus not on ways to employ or combine concepts, as we do in science or everyday life, but rather on recovering what we meant in the first place by our most fundamental acts of thought. This is a difficult, painful process that involves a redirection of our thinking toward self-reflection. Since both Gödel and Husserl (in his later period) were conceptual realists, the self-reflection at issue concerns understanding how we grasp real, objective concepts; so the subjectivism Goldfarb feared was an illusion. Subjectivism, as Goldfarb employed this term, is in opposition to objectivism. What Gödel found valuable in Husserl, however, was a turn to the thinking subject, the source of cognition, which was meant not as an alternative to objectivism, but rather as an account of how what is objective is given to us. Indeed, "in the last analysis," wrote Gödel, "the Kantian philosophy rests on the idea of phenomenology, albeit in a not entirely clear way." And it is phenomenology, according to Gödel, which "entirely as intended by Kant, avoids both the death-defying leaps of idealism into a new metaphysics as well as the positivistic rejection of all metaphysics."

From Frege to Gödel

Goldfarb's invocation of Frege was especially misplaced. To begin with, the review by Frege he cited concerned Husserl's early, non-Platonist, psychologistic account of arithmetic, whereas the phenomenology that caught Gödel's eye was an attempt by Husserl late in his career to reconcile mathematical Platonism and conceptual realism with the capacities of the human intellect. Frege's early critique does not speak to this project.

Indeed, one can ask how exactly Frege came up with his own analysis of the concept of number in his path-breaking study *The Foundations of Arithmetic*. A natural number, Frege begins, is what answers the question "How many?" ("How much?" is answered by a real number.) But how many *what*? First we must determine what it is that is really numbered. His answer: not the nine planets themselves (each by itself is one, and the group as such is also one), but rather the concept *is a planet*. To assign a number to a concept, then, is to determine how many instances it has. The number nine numbers the planets because the concept *is a planet* has nine instances. Such are the opening moves in Frege's deep and beautiful analysis of the concept of number, at the conclusion of which he offers an explicit definition: the number assigned to the concept F is the extension of the concept *equinumerous with F*. (Concepts are equinumerous when there is a one-to-one correlation between their extensions, i.e., to the set of objects falling under each. In effect, then, for Frege, the number three is the set of all trios, i.e., of all things that are three in number.) Nothing like this had ever been proposed before. As Frege himself put it, his discussion raised the subject to an entirely new level.

But where did this profound analysis come from? Was it just a lucky guess? Was it the result of an empirical survey Frege had conducted among the citizens of Jena? Or perhaps the great logician was simply grinding out theorems from some preexisting set of axioms? None of these "explanations," of course, holds water, nor was Frege simply sitting back and regarding the "passing show." He had turned his attention, as no one had before him, to what we are really doing when we employ the concept of number, when we use numbers, that is, both to count things and to do arithmetic. But this redirection of attention is just what Gödel, following Husserl, called phenomenology. And Wittgenstein, too, it would appear. At one point he remarked to a friend that one could with justice call his method of investigation of the correct use of words, phenomenology. Where Wittgenstein speaks of words, Gödel refers to concepts. "'Trying to see (i.e., understand) a concept more clearly,'" says Gödel, "is the correct way of expressing the phenomenon vaguely described as 'examining what we mean by a word.'"

The goal of this method is to enable us to discern the concept itself, free from the encrustations of historical practice and psychological necessity. This was as much Frege's goal as it was Gödel's. As Frege put it, "Often it is only after immense intellectual effort, which may have been continued over centuries, that humanity at last succeeds in achieving knowledge of a concept in its pure form, in stripping off the irrelevant accretions which veil it from the eyes of the mind." But concepts, of course, aren't physical objects. We can't literally see them. Rather, says Gödel, "we perceive objects and understand concepts. Understanding is a different kind of perception." He adopted the Kantian term "intuition" for the quasi-perceptive grasping by the mind's eye of concepts and other "ideal objects." Goldfarb, once again, is offended: "Kantian 'intuition' is a matter of the presentation of objects (in sensibility) in space and time; none of this is in

Gödel's notion." Why then did Gödel continue to employ this term for his own purposes?

Goldfarb makes no effort to resolve this question. Why not? For the question does have an answer, which lies in the fact that for a Platonist or conceptual realist, the mind encounters an ideal entity in a manner parallel to the way the eye tracks a physical object. In both cases, we are confronted with something real that we have not ourselves created. We grasp it, therefore, only partially and in stages, gaining new insights as we shift our perspective. "We begin with vague perceptions of a concept," says Gödel, "as we see an animal from far away or take two stars for one before using the telescope." Since the entity is not a child of our own imagination, we will never exhaust the information to be gained by different ways of approaching it, but we may reach a limit after which we no longer find ourselves bumping up against surprises or running into mysteries. If that threshold is achieved—as Gödel thought it was by Turing's analysis of the concept of effective computability—we are satisfied that our perception, or quasi-perception, is now adequate to our purpose and that we have an acceptable (though not perfect) intuition of the concept or object. This explains why Gödel rejected Kant's more limited conception of intuition, as well as why he chose to retain the Kantian term while giving it a more generous interpretation.

Goldfarb's attempt to enlist Frege in his critique of Gödel was worse than misguided. It kept him from noting the deep philosophical affinities between these two giants of modern logic. They swam in a sea of ultraempiricism, but managed somehow not to get wet. Their philosophy was an uncompromising Platonism, reminiscent of Plato himself. "In arithmetic," wrote Frege, "we are not concerned with objects which we come to know as something alien from without through the medium of the senses, but with objects given directly to our reason, and, as its nearest kin, utterly transparent to it." Plato could not have put it better. As he wrote in the *Phaedo*, "When the soul investigates by itself it passes into the realm of what is pure, ever existing . . . and unchanging . . . [and is] akin to it."

Frege and Gödel were "logicists" who believed that mathematics, with the exception of geometry, could be derived from logic, which they took to include set theory, together with the correct definitions of the fundamental concepts. Neither believed, as did the positivists, that the derivability from definitions, or analyticity, of arithmetic meant that its formulas lacked cognitive content. Gödel made it clear that what is analytic is true in virtue of the meanings of the concepts involved. Each held that geometrical knowledge was different, founded on a kind of a priori physical intuition, as intimated by Kant, a philosopher both had studied deeply, though they rejected large parts of his thinking. Each believed that the fundamental logical and metaphysical relationship is that of predication—the theory of which, predicate logic, Frege had invented from whole cloth—whose metaphysical correlative is falling under a concept (the central idea, not accidentally, of Frege's philosophy of arithmetic). Both held that the only means of escaping from the personal, solipsistic world of the ideas in your head was to grasp abstract concepts, which exist not in our minds but rather in an objectively real "concept space," and that by this means we are granted access both to the physical world (by employing the concept of physical object) and to what Frege called the "third realm" of ideal entities. "Concepts are there," wrote Gödel, "but not in any definite place. They . . . form the 'conceptual space,'" while for Frege, "in the external world, in the whole of space and all that therein is, there are no concepts . . . no numbers."

One place where Gödel went beyond Frege was in attempting to forge a systematic epistemology that would account for our ability to make contact with "conceptual space" or the "third realm." Here Gödel turned to Husserl, whom he saw as improving on the Kantian philosophy, which is strong in epistemology but weak in ontology (weak, that is, for realists like Gödel, Frege and Husserl). Yet perversely, it is just here, where a serious Platonist has finally decided to settle his epistemological accounts, that Goldfarb chooses to take Gödel to task—attempting to enlist Frege against Gödel just when Gödel was engaged in repairing a hole in Frege's philosophy—by citing

Frege's irrelevant critique of Husserl's earlier, non-Platonist work. And this from a fellow logician, indeed, one of the guardians of Gödel's legacy, an editor and contributor to the definitive *Collected Works* of Kurt Gödel. How is this to be explained?

Goldfarb and Dreben's performances at the Boston University conference (and in Goldfarb's case, also at Helsinki) were no aberration. In their steadfast refusal to find anything of value in Gödel's contributions to philosophy and their unembarrassed display of condescension, they were simply carrying on a great and noble tradition among professional philosophers. Long before the two spoke in Boston, for example, Charles Chihara, a philosopher of mathematics at the University of California at Berkeley, had written a series of essays in which he took pains not just to criticize but to ridicule Gödel's mathematical epistemology. He continued this project in a book published in 1990, *Constructibility and Mathematical Existence*, which proved too much for at least one reviewer. "Unfortunately," wrote E.P. James, "Chihara's account of Gödelian Platonism follows the common line of regarding Gödel as a logician par excellence but a philosophical fool." James went on to demonstrate that Chihara had simply gotten Gödel wrong, concluding that "having failed to appreciate the complexity of Gödel's beliefs about physical perception, it is easy for Chihara to argue against a caricature of his analogous beliefs about mathematical 'perception.'" Once again, Gödel's invocation of mathematical intuition had succeeded in arousing the ire of a professional philosopher. "At present," Gödel wrote, "mathematicians are prejudiced against intuition." Philosophers, too. For Gödel's philosophy of time, on this point as on others, has been accorded a similarly cold reception.

"The correct response to Gödel"

In his 1987 book *Asymmetries in Time*, Paul Horwich, then at MIT, offered fresh reexaminations of such staples of the space-time literature as the "direction of time" and the difference between past and future. One aspect that was new was his decision to revisit Gödel's argument for the possibility of time travel, and newer still was the conclusion he reached, that the argument was actually valid. What was not new was his neglect of the philosophical motive—the demonstration of the ideality of time—behind Gödel's development of his new world models. In contrast, Milič Čapek, another philosopher of science, had argued decades before Horwich against Gödel's case for time travel precisely on the grounds that if it held water, it followed that time was not real, a conclusion he found unacceptable. (Karl Popper had argued similarly.) Thus, an ironic situation: Čapek (but not Horwich) agreed with Gödel that the possibility of time travel signaled the ideality of time, while Horwich (but not Čapek) took sides with Gödel on the genuine possibility of time travel. Neither sided with Gödel in maintaining both that time travel is possible and that therefore time is merely ideal.

Attention, however, was paid to this dual aspect of Gödel's reasoning in a small book of mine published in 1991, *The Disappearance of Time: Kurt Gödel and the Idealistic Tradition in Philosophy*. When an expanded edition appeared in 1999, some notice was taken, and a small but growing cottage industry developed of attempts to assess what Gödel was really up to in his writings on Einstein. The zeitgeist, however, is not so easily deflected. On the question of whether Gödel really had something important to contribute to the philosophy of time, as well as to the philosophy of mathematics and the ontological argument for God, an impolite skepticism remains dominant.

That the spirit of the time is still unmoved became most apparent in 1995, the year of the Gödel conference in Boston, when a distinguished philosopher of space and time, John Earman, responded to me and others by devoting an appendix of his book *Bangs, Crunches, Whimpers, and Shrieks* to Gödel's argument for the ideality of time. He began auspiciously by noting "the relative neglect of the philosophical moral Gödel himself wanted to draw from his solutions to EFE [Einstein's field equations]." He continued in like manner by observing, with becoming piety, that "a deeply held conviction of someone of Gödel's stature deserves serious consideration." At long last, it appeared, the space-time establishment was giving Gödel his due. But the appearance, like time itself, was an illusion. Earman's assessment of what Gödel had to teach us was entirely negative. He couldn't resist adding, moreover, that the neglect of Gödel's philosophical moral had after all been "benign." Then, just in case his message wasn't sufficiently clear, he concluded by observing that Einstein, in his reply to Gödel in the Schilpp volume, had "brushed aside" Gödel's discussion of idealism, adding that "this seems to me to be the correct response to Gödel." So much for the attention due someone of Gödel's stature.

That Earman took issue with Gödel is not the point. The point is that once again a noted philosopher felt no embarrassment in dismissing Gödel with maximum condescension. Moreover, just as Goldfarb had attempted to enlist someone Gödel admired, namely Frege, to attack him, Earman employed Einstein to back up his claim that Gödel's attempt at philosophy should be "brushed aside." Once again, however, the enlistment was premature. It is true that Einstein, in his response to Gödel, sidestepped the issue of idealism, but he was speaking as a physicist whose interest was in the implications of Gödel's important new discovery (as Einstein described it) for the theory of relativity. He was not, like Goldfarb, participating in a conference devoted to Gödel's contributions to philosophy, nor, like Earman, discoursing on the philosophy of time for an audience of philosophers. On his own, Einstein shared Gödel's appreciation of Kant, and also his reservations, and he too had occasionally speculated on the relationship between relativity theory and our everyday experience of time. Unlike Gödel, however, he had no intention of engaging directly with the philosophical literature, and had no plans to follow up his philosophical speculations in any depth. He was not, in short, an appropriate ally in a campaign to put Gödel in his philosophical place.

Einstein aside, Earman made it clear that nothing in Gödel's argument impressed him. He looked askance at everything, even something as straightforward as Gödel's clear appreciation that if time flows, this flux represents the coming into existence of new states of affairs, and that the relativity of simultaneity could not sensibly be taken to imply the relativity of existence. Without disturbing himself to dispute Gödel's conclusion, Earman commented only that "this is a pretty piece of ordinary language philosophizing . . . but like most of its ilk, it leaves one up in the air . . . one can wonder how such intuitions can support such weighty philosophical morals." Once again Gödel's invocation of intuition had gotten him into trouble. But Earman had added a new twist. Somehow, Gödel, with his distrust of "ordinary language philosophizing" exemplified by Wittgenstein and his school, had been lumped together with this movement. The intuitions Gödel had in mind, however, were the result of the highly rational exercise of turning one's attention to the nature of the concepts themselves, not of tuning one's ear to the marketplace of "ordinary language" and everyday conversation. Language as such, for Gödel, had nothing to do with it.

The crux of Earman's critique concerned the relevance of time travel in a merely possible Gödel universe—where one might grant that time is an illusion—to the existence of time in our own world. Since the extreme geometry of the Gödel universe, which allows for the possibility of closed, future-directed, timelike curves, is not a feature of the actual world, Earman argued, Gödel had failed to show time's absence in the world in which we live. Yet Earman did not deny that the experience of the lapse of time does not decide the issue, since it is by hypothesis something we might well have in common with the denizens of the Gödel universe. That the geometry of our world, moreover, does not in itself exclude the possibility of the flow of time is also not decisive, since that shows at most that the conditions necessary for the existence of time are present in our world. But necessary conditions are not the same as sufficient conditions, and absent the testimony of the experience of time (which cuts both ways), and of the laws of nature (which are the same in both worlds), it is hard to see that anything at all could decide in favor of the existence of time in the actual world. In short, our world would appear, by Gödel's lights, to be in principle indistinguishable from a universe in which time is demonstrably absent.

How could one miss the force of this argument? In company with other philosophers of physics, Earman verges on conflating the geometrical prerequisites for the lapse of time with the actual lapse itself. Indeed, the latter is something Earman clearly has trouble taking seriously. But if we don't acknowledge the flow of time, one can hear Gödel saying, then the game is already over: there simply is no such thing as time in the intuitive sense, and an argument for the ideality of time is not needed. In fact, Gödel actually did say just this: "One may take the standpoint," he wrote in the Schilpp volume on Einstein, "that the idea of an objective lapse of time (whose essence is that only the present really exists) is meaningless. But this is no way out of the dilemma; for by this very opinion one would take the idealistic viewpoint."

One can, of course, in good conscience, disagree with Gödel's conclusions. He did not suffer from the illusion that he had said the last word on the reality of time, which, as we saw earlier, he described to Hao Wang as perhaps *the* philosophical question. But it is difficult to understand how one can disagree with the judgment that, whether he is ultimately right or wrong, Gödel has made a profound contribution to philosophy. In addressing the question of time's existence, a question that has haunted philosophers from Parmenides and Plato to Kant, he brought to bear the most powerful, most fully developed formal account of time ever constructed, the theory of relativity, and proceeded to distinguish, with great precision, the formal from the intuitive concept, relativistic from intuitive time. He then demonstrated, simply and elegantly, that the existence of intuitive time, which lapses, is inconsistent with the truth of special relativity, in particular, with the relativity of simultaneity and the equivalence of all inertial frames. He proceeded, next, to remind us that in general relativity certain reference frames can be seen as privileged, and that—if the distribution of matter and motion is accommodating—we can construct a simulacrum of intuitive time, namely cosmic time. He then made a deep and surprising contribution to relativistic physics by discovering new world models in which, due to the unhappy distribution of matter and motion, even cosmic time cannot be constructed. In addition, he proved that in some of these new world models, or Gödel universes, time travel is possible, which once again proves that in such worlds there is no such thing as cosmic or intuitive time. To clinch the deal, he brought to bear a modal argument, from possibility to actuality, making the case that if time was ideal in the Gödel universe, which contains the same physical laws as ours, and whose inhabitants might well experience time in the same way we do, then it cannot be that the mere difference in the distribution of matter and motion in the two worlds accounts for the fact that in one time exists, whereas in the other it does not. Finally, in a deep and original interpretation of Kant, who is usually taken to have been refuted by Einstein, he argued that, correctly understood, Kant's doctrine of the ideality of time bears striking affinities with the temporal idealism implicit in the theory of relativity. If this is not the right way to do philosophy, what is?

Gödel as Philosopher

The case for Gödel as philosopher is unassailable. Though he published few essays that could be considered explicit contributions to philosophy, they suffice to establish him as an important philosopher of mathematics and of space and time. The posthumous publication of several more of his philosophical studies, including the Gibbs Lecture, his contribution to the Schilpp volume on Carnap, and the longer version of his contribution to the Schilpp volume on Einstein, confirm this assessment. The essay he wrote for the Schilpp volume on Russell, which contained new and insightful discussions of Frege as well as Russell on the question of meaning, including an illuminating and prescient comparison of Russell on "denoting" with Frege on "sense and reference," leaves little room for doubt that in the philosophy of language, too, his abilities were striking. He had clearly mastered the writings of most of the seminal figures in twentieth-century analytical philosophy, including Frege, Russell and Carnap.

He knew Wittgenstein's *Tractatus*, too, though it is not known how well, and was a student of the writings of the founding father of continental philosophy, Husserl. The split between the analytical and continental schools, which, sadly, holds sway to this day, did not intimidate him. Here as elsewhere he proved himself free from the philosophical prejudices around him. He attended seriously, as well, to the history of philosophy, devoting endless hours to the study of Leibniz and acquiring a profound understanding of Kant. His grasp of Hegel astonished the logician and philosopher Georg Kreisel, a man not easy to impress. Looking over the set of quotations from Hegel that Gödel had assembled, Kreisel remarked that "the publication of such an anthology is likely to produce a minor revolution in philosophy." He also studied Plato and Aristotle, as well as the medieval philosophers, but we do not know the extent of his familiarity with these figures. We do know that he put his grasp of the history of philosophy to creative use, enlisting his knowledge of Kant to help him comprehend the philosophical significance of the theory of relativity, and turning to Husserl's phenomenology for assistance in developing an epistemology adequate to the Platonist ontology he espoused for mathematics. He believed that the history of philosophy could help free us from prejudice. "Even science," he said, "is very heavily prejudiced in one direction. Knowledge in everyday life is also prejudiced. Two methods to transcend such prejudices are: (1) phenomenology; (2) going back to other ages."

Overarching much of his research in philosophy and logic was the "Gödel program," the investigation of the limits of formal methods in capturing intuitive concepts. This was clearly a philosophical enterprise, though he carried it out using both formal and philosophical tools. If the leitmotif of the twentieth century was formalism, in the most general sense, his incompleteness theorem was unquestionably the single most important contribution to this subject. As it had been for Plato, mathematics was for him a deep source of philosophical inspiration, in itself, in its relationship to logic, and in its ability to describe the physical world. A lifelong opponent of positivism, both in the narrow, technical sense of the program embraced by the Vienna Circle and in the broader sense of a philosophical tendency endemic to every age, he attempted to reappropriate mathematics and logic for the other side. "One bad effect of logical positivism," he said in his conversations with Wang, "is its claim of being intimately associated with mathematical logic. . . . Mathematical logic should be used more by nonpositivistic philosophers. The positivists have a tendency to represent their philosophy as a consequence of logic, to give it scientific dignity."

If in his published writings he aimed for maximum precision and minimum controversy, stripping down his contributions until only the bones were left—that is, until all that remained was what was most amenable to rigorous demonstration and unavoidable philosophical interpretation—in his notebooks and in his conversations with Wang, he felt free to engage in the most wide-ranging, fundamental speculations, flying through thin air as high as pure thought could take him, with no fear of crashing since he had no intention of landing. There is a fair amount of rough coal in these extravagant musings, but there are also diamonds, sparks of insight into noncausal "laws" of historical development, the limitations of mechanical biology, a concept of "absolute" proof, and the possibility of an afterlife. If Einstein accepted his end with equanimity, understanding that he had long since fired off his best shots, it is understandable that Gödel resented his. In many ways, he had just begun to fight.

An Ugly Body in Beautiful Clothing

Even on his deathbed, Einstein had been engaged in a desperate effort to find a unified field theory. When his eyes finally closed, it fell to Bruria Kaufman, his last collaborator, and his friend Gödel to attend to the papers that remained in the great physicist's office. Together, they found a blackboard filled with sad equations that led nowhere. The fruits of Einstein's earlier endeavors, however, remain the jewels in the crown of physics in the twentieth century. His theories uncovered the mathematical symmetries hidden in the noise and confusion of the physical world.

This fact deeply impressed Gödel. Mathematics itself, which he had encountered more directly than Einstein had, was for him a source of wonder and admiration. "It is given to us in its entirety and does not change," he said, "unlike the Milky Way. That part of it of which we have a perfect view seems beautiful, suggesting harmony." More than this, however, he recognized that these symmetries are not exclusive to the separate world of pure, transcendent "forms." "Mathematics is applied to the real world," he wrote, "and has proved fruitful. This suggests that the mathematical and empirical parts are in harmony and that the real world is also beautiful." Like Plato in the *Timaeus*, he believed that however the "real" world had come about, it was based on a divine model. "Otherwise," he said, "mathematics would be just an ornament and the real world would be like an ugly body in beautiful clothing."

On his departure from this world, he left boxes of papers at the institute, begging for discovery. The first to do so, his biographer John Dawson, found two bound notebooks containing calculations that had nothing to do with logic or pure mathematics. It turned out they were Gödel's recordings of the angular orientations of galaxies. The purist of pure logicians had never ceased trying to discover whether the actual world we inhabit is a Gödel universe. His efforts, however, were unnecessary. In a deep sense, it is clear enough that we all do live in Gödel's universe.

Notes

Chapter 1

2 **key to his trunk:** See Peter Suber, 1992, "Fifty Years Later, the Questions Remain: Kurt Gödel at Blue Hill," *Ellsworth American*, August 27.

2 **"From a distance":** Albrecht Fölsing, 1998, *Albert Einstein*, trans. Ewald Osers (New York: Penguin Books), 689.

2 **"just to have the privilege":** Oskar Morgenstern, in a letter to the Austrian government in 1965, cited in Hao Wang, 1996, *A Logical Journey: From Gödel to Philosophy* (Cambridge, Mass.: MIT Press), 57.

3 **"the most significant":** W.V. Quine, 1952, text of citation for Gödel's honorary doctorate from Harvard University, June, cited in Wang 1996, 2.

4 **"Here you see that portion":** Fölsing 1998, 283.

4 **"When I objected":** Werner Heisenberg, 1983, *Encounters with Einstein and Other Essays on People, Places, and Particles* (Princeton: Princeton University Press), 114.

4 **"on equal terms with Einstein":** Freeman Dyson, 1993, *From Eros to Gaia* (New York: Penguin Books), 161.

5 **"Only fables":** Gödel, letter to his mother, April 21, 1965, cited in Wang 1996, 45.

5 **"You know, once you start calculating":** Fölsing 1998, 311.

6 **"Every boy in the streets":** Constance Reid, 1986, *Hilbert–Courant* (New York: Springer-Verlag), 142.

7 **Time, "that mysterious. . .":** Kurt Gödel, "A Remark About the Relationship Between Relativity Theory and Idealistic Philosophy," in Paul Arthur Schilpp, ed., 1949, *Albert Einstein: Philosopher-Scientist* (LaSalle, Ill.: Open Court), 557.

8 **"chronology protection conjecture":** Stephen Hawking, 1992, "Chronology Protection Conjecture," *Physical Review* D46, no. 2 (July 15).

Chapter 2

9 **The German man of science:** Arthur Miller, 1986, *Imagery in Scientific Thought* (Cambridge, Mass.: MIT Press), 108.

9 **It is a remarkable fact:** Kurt Gödel, 1946–1949, "Some Observations About the Relationship Between Theory of Relativity and Kantian Philosophy," in Solomon

Feferman et al., eds., 1995, *Kurt Gödel: Collected Works*, vol. 3, *Unpublished Essays and Lectures* (New York: Oxford University Press), 230.

10   **"I have firmly resolved"**: Fölsing 1998, 333.

10   **"Of course he has no children"**: Suber 1992.

10   **"There is no doubt"**: Paul Weingartner and Leopold Schmetterer, 1987, eds., *Gödel Remembered* (Naples: Bibliopolis), 32.

10   **"Albert himself"**: Dennis Overbye, 1999, "Einstein, Confused in Love and, Sometimes, Physics," *New York Times*, August 31.

11   **Finding himself trapped**: Paul Halmos, 1988, *I Want to Be a Mathematician: An Automathography in Three Parts* (Washington, D.C., Mathematical Association of America).

11   **"to make the serious things in the world tolerable"**: Philipp Frank, 1989, *Einstein: His Life and Times* (New York, Da Capo Press), 281–282.

12   **Einstein "was an experienced sight reader"**: Gerald Holton and Yehuda Elkana, 1982, eds., *Albert Einstein: Historical and Cultural Perspectives* (Mineola, N.Y.: Dover), 410.

12   *Begriffsschrift*: Gottlob Frege, 1967, *Begriffsschrift*, A Formula Language, Modeled upon That of Arithmetic, for Pure Thought," in Jean van Heijenoort, ed., *From Frege to Gödel: A Source Book in Mathematical Logic, 1879–1931*, 182 (Cambridge, Mass.: Harvard University Press).

12   **"deeply religious unbeliever"**: Gerald Holton, 1998, "Einstein and the Cultural Roots of Modern Science," *Daedalus*, Winter, 16.

13   **Gödel was not a pantheist**: Hao Wang, 1987, *Reflections on Kurt Gödel* (Cambridge, Mass.: MIT Press), 18.

13   **Spinoza's God:** Wang 1996, 152.

13   **Whereas "ninety per cent of philosophers"**: Ibid., 107.

13   **"All three of the others"**: Bertrand Russell, 2000, *Autobiography* (London: Routledge), 466.

13   **"I am not a Jew"**: Wang 1996, 112.

13   **"Kurt had a friendly attitude"**: Weingartner and Schmetterer 1987, 33.

14   **"I frequently held an opinion"**: John Dawson, 1997, *Logical Dilemmas* (Wellesley, Mass.: A.K. Peters).

15   **"You became a mathematician"**: Wang 1996, 87.

15   **"You have a vicarious"**: Karl Menger, 1994, *Reminiscences of the Vienna Circle and the Mathematical Colloquium* (Dordrecht, Netherlands: Kluwer Academic Publishers), 222.

15   **"Who ever became more intelligent"**: Ibid., 223

15   **"If you had a particular problem"**: Weingartner and Schmetterer 1987, 33.

16   **"Kant is a sort of highway"**: Ibid., 74.

16   **"At the Institute in Princeton"**: Holton 1998, 18.

18   **"holy geometry booklet"**: Albert Einstein, "Autobiographical Notes," in Schilpp 1949, 11.

18   **"Time," Kant himself said:** Norman Kemp Smith, 1965, trans., *Critique of Pure Reason*, by Immanuel Kant (New York: St. Martin's Press), B 50.

18   **"We cannot obtain for ourselves"**: Ibid., B 156.

18   **"the primary object of Einstein's"**: G.J. Whitrow, 1980, *The Natural Philosophy of Time* (Oxford: Clarendon Press), 4.

Chapter 3

21 "After one session": Menger, 1994, 210.

22 "You are me!": Ray Monk, 1990, *Ludwig Wittgenstein: The Duty of Genius* (New York: Free Press), 108.

22 class photograph: See the book jacket of Kimberley Cornish, 1998, *The Jew of Linz* (London: Century).

23 "God made the natural numbers": Kronecker, quoted in E.T. Bell, 1986, *Men of Mathematics* (New York: Touchstone Books), 477.

25 *Principia Mathematica:* Bertrand Russell and A.N. Whitehead, 1910–1913, *Principia Mathematica* (Cambridge, Cambridge University Press).

25 *Introduction to Mathematical Philosophy:* Bertrand Russell, 1919 (London: Allen & Unwin).

26 Gödel "always grasped problematic points": Menger 1994, 205.

26 Olga Taussky-Todd's account: Weingartner and Schmetterer 1987, 32.

27 but possessing "a real flair": Georg Kreisel, 1980, "Kurt Gödel: 1906–1978," *Biographical Memoirs of Fellows of the Royal Society* 26:154.

27 by constructing "far-fetched grounds for jealousy": Ibid, 154.

28 *Tractatus*: Ludwig Wittgenstein, 1961, *Tractatus Logico-Philosophicus*, trans. D.F. Pears and B.F. McGuinness (London: Routledge & Kegan Paul).

28 "what we cannot speak about": Wittgenstein 1961, 7

28 "Today we . . . out-Wittgensteined these Wittgensteinians": Menger 1994, 210.

30 "I . . . believe myself to have found": Wittgenstein 1961, 5.

30 "I can testify to this": Weingartner and Schmetterer 1987, 40.

32 "I put my faith in organization": Holton 1998, 16.

32 "Despite their remoteness": Kurt Gödel, 1964, "What Is Cantor's Continuum Problem?" (1964 supplement), in Feferman et al. 1990, II/268.

33 "Mach's way": Holton 1998, 9.

34 "since the mathematicians pounced on the relativity theory": Fölsing 1998, 245.

35 "The antipathy of these scholars": Albert Einstein, "Autobiographical Notes," in Schilpp 1949, 49.

35 "We really experience only": Gottlob Frege, "Thought," in Michael Beancy, ed., 1997, *The Frege Reader* (Oxford: Blackwell), 339.

35 Lenin's critique of Mach: V.I. Lenin, 1950, *Materialism and Empirio-Criticism*, trans. A. Fineberg (London: Lawrence & Wisehart).

37 "I don't believe that atoms exist!": David Lindley, 2001, *Boltzmann's Atom* (New York: Free Press), vii, xi.

37 Einstein's *Annalen der Physik* publications: Albert Einstein, 1905a, "On the Electrodynamics of Moving Bodies," *Annalen der Physik* 17:891–921; Albert Einstein, 1905b, "On the Movement of Particles Suspended in Fluids at Rest, as Postulated by the Molecular Theory of Heat," *Annalen der Physik* 17:549–60.

38 "the scientist must appear": Schilpp 1949, 684.

38 Wittgenstein's view of "logical space": Wittgenstein 1961, 1.13 and 2.11.

39 His most influential work: Moritz Schlick, 1918, *Allgemeine Erkenntnislehre* (Berlin: Springer-Verlag).

39 "Sad, but true": Menger 1994, 56.

40 "Now you damned bastard": See David Edmonds and John Eidinow, 2001, *Wittgenstein's Poker* (New York: HarperCollins), 142–148.

41 "It is to be hoped": Ibid., 146.

41 Nelböck after the Anschluss: Ibid., 147.

47 "Dear Colleague": Bertrand Russell, 1902, "Letter to Frege," in van Heijenoort 1967, 124.

48 "The sole possible foundations of arithmetic": Frege, 1902, "Letter to Russell," in van Heijenoort 1967, 128.

48 "The development of philosophy since the Renaissance": Kurt Gödel, 1961, "The Modern Development of the Foundations of Mathematics in the Light of Philosophy," in Feferman et al. 1995, III/375–376.

49 "Thus came into being": Ibid., 379.

49 in Gödel's words: Ibid.

49 "an a priorism with the sign reversed": Ibid., 383.

49 "[A] system of truths": Warren Goldfarb, "The Philosophy of Mathematics in Early Positivism," in R.N. Giere, ed., 1996, *Origins of Logical Empiricism*, Minnesota Studies in the Philosophy of Science, vol. 16, 213–230 (University of Minnesota Press), 214.

49 "the unreasonable effectiveness of mathematics": Eugene Wigner, 1979, "The Unreasonable Effectiveness of Mathematics in the Natural Sciences," in *Symmetries and Reflections: Scientific Essays*, 222–237 (Woodbridge, Conn.: Ox Bow Press).

50 "It is not true, as Kant urged": Hans Hahn, "The Crisis in Intuition," in Brian McGuinness, ed., 1980, *Empiricism, Logic, and Mathematics: Philosophical Papers,* 731–02 (Dordrecht, Netherlands: D. Reidel), 101.

50 "The essence of this view": Kurt Gödel, 1951, "Some Basic Theorems on the Foundations of Mathematics and Their Implications" (Gibbs Lecture, 1951), in Feferman et al. 1995, III/319.

Chapter 4

51 Every spy's life: Anaïs Nin, 1974, *A Spy in the House of Love*, (Athens, Ohio: Swallow Press), 73.

51 "[Russell] brought to light": Kurt Gödel, 1944, "Russell's Mathematical Logic," in Feferman et al. 1990, II/124.

52 "Where else would reliability and truth be found": David Hilbert, 1925, "On the Infinite," in van Heijenoort 1967, 375.

52 "The definitive clarification": Ibid. 370–371.

52 "As concerns the world": John Dawson, 1985, "Discussion on the Foundations of Mathematics," *History and Philosophy of Logic* 5:116.

52 "A very simple fact now seems": Ibid.

53 Who pays any attention: e. e. cummings, "since feeling is first," 1926, in *Complete Poems: 1904–1962* (New York: Liveright, 1994).

54 "I can begin to hear the sound of machinery": Ray Monk, 1990, *Ludwig Wittgenstein: The Duty of Genius* (New York: Free Press), 13.

55 Frege's rejection of implicit definition: Gottlob Frege, 1980, *The Foundations of Arithmetic: A Logico-Arithmetic Enquiry into the Concept of Number*, trans. J.L. Austin (Evanston, Ill.: Northwestern University Press), 68.

57   "Continued appeals to mathematical intuition": Kurt Gödel, "What Is Cantor's Continuum Problem?" (1964 supplement), in Feferman et al. 1990, II/269.

58   "I would be very much interested": Dawson 1997, 70.

58   "Von Neumann, from the beginning": Dawson 1984, 122.

58   Von Neumann's striking prescience: S.M. Ulam, 1976, *Adventures of a Mathematician* (Berkeley: University of California).

62   "I was rather depressed": Benjamin Yandell, 2002, *The Honors Class: Hilbert's Problems and Their Solvers* (Natick, Mass.: A.K. Peters), 75.

68   Similarly, as a direct consequence: William F. Dowling, 1989, "There are no safe virus tests," *American Mathematical Monthly* 96, 835–836.

69   God's mercy preserves mathematics: Simone Weil, 1987, "The Pythagorean Doctrine," in *Intimations of Christianity Among the Ancient Greeks* (London: Ark Paperbacks), 165.

71   Hilbert and Grommer: Yandell 2002, 18.

72   Hilbert and Max Born: Reid 1986, 105.

72   "Formalists considered": Kurt Gödel, letter to Hao Wang, 1968, in Feferman et al. 2003, V/404.

73   Gödel and Zermelo at Bad Elster: Dawson 1997, 76.

73   Russell's comments: Ibid., 77.

73   "Russell evidently misinterprets my result": Kurt Gödel, 1973, letter to Abraham Robinson, in Feferman et al. 2003, V/201.

74   Post noted: Emil Post, "Absolutely Unsolvable Problems and Relatively Undecidable Propositions: Account of an Anticipation," in Martin Davis, ed., 1965, *The Undecidable: Basic Papers on Undecidable Propositions, Unsolvable Problems and Computable Functions*, 340–433 (New York: Raven Press), 345.

75   we have "for the first time succeeded . . .": Kurt Gödel, 1946, "Remarks before the Princeton Bicentennial Conference on Problems in Mathematics," in Feferman et al. 1990, II/150.

Chapter 5

77   "In the midst of the exultant joy": Walter Moore, 1989, *Schrödinger: Life and Thought* (Cambridge: Cambridge University Press), 334.

78   "I am fire and flame for it": Ed Regis, *Who Got Einstein's Office? Eccentricity and Genius at the Institute for Advanced Study*, Reading, Mass., Addison-Wesley, 1987, 22.

80   "When one returns again": Walter Moore 1989, 334.

81   Of course he had known: Ibid., 338.

82   "I get buckets and buckets": David Edmonds and John Eidinow, 2001, *Wittgenstein's Poker* (New York: HarperCollins), 127.

83   "Can you name me a Jew": Fred Prieberg, 1991, *Trial of Strength: Wilhelm Furtwängler and the Third Reich*, trans. Christopher Dolan (Quartet Books), 94.

83   "I . . . felt discouraged": Miller 1986, 143.

83   "the more I reflect": Ibid.

84   (This neglect has begun to be corrected.): See Thomas Levenson, 2003, *Einstein in Berlin* (New York: Bantam Books).

84   "It is not entirely clear": Fölsing 1998, 330.

87   "I am told in all steamship bureaus": Dawson 1997, 149.

88   "das Reich der Zwei": Kurt Vonnegut, 1999, *Mother Night* (New York: Delta Books).

Chapter 6

89   When pygmies cast such long shadows: Gian-Carlo Rota, 1997, *Indiscrete Thoughts* (Boston: Birkhaüser), 257.

89   "as like Oxford as monkeys can make it": Ronald W. Clark, 1976, *The Life of Bertrand Russell*, Alfred A. Knopf, New York, 233.

89   "Princeton is a wonderful piece of earth": Fölsing 1998, 679.

89   "ten times more congenial": Dawson 1997, 168.

91   "there is any danger": Ibid., 158.

92   "Is his [mental] illness": Yandell 2002, 52.

93   "in today's terms, Einstein": Peter Bucky, 1992, *The Private Albert Einstein* (Kansas City: Andrews and McMeel), 102.

93   Feynman's "physiognomy and manner": James Gleick, 1992, *Genius: The Life and Science of Richard Feynman* (New York: Pantheon Books), 84.

94   "during my sickness Einstein": Hao Wang, 1987, *Reflections on Kurt Gödel* (Cambridge, Mass.: MIT Press), 37.

94   Gödel wrote to his mother: Wang 1996, 39.

94   "I know of one occasion": Wang 1987, 32.

95   "That begins only": Wang 1996, 56.

95   Einstein . . . "was in many respects a pessimist": Wang 1987, 39–40.

95   "Gödel," however: Ibid., 32.

95   "museum pieces": Holton 1998, 4.

96   "Einstein is completely cuckoo": Fölsing 1998, 693.

96   "firing at birds in the dark": Ibid., 709.

98   "Did I really say that?": Ibid., 716.

98   "Cervantes' text": Jorge Luis Borges, 1964, "Pierre Menard, Author of the Quixote," in *Labyrinths: Selected Stories and Other Writings* (New York: New Directions), 42.

99   "Do you think a dictatorship": Dawson 1997, 180.

100   "You know that I disagree": Mark van Atten and Juliette Kennedy, 2003, "Gödel's Philosophical Development," *The Bulletin of Symbolic Logic*, vol. 9, no. 4, December, 425–476, 470.

100   In 1935, Einstein: Albert Einstein, Boris Podolsky, and Nathan Rosen, 1935, "Can Quantum-Mechanical Description of Physical Reality Be Considered Complete?" *Physical Review* 47 (May 15), 777–780.

101   "The heuristics of Einstein and Bohr": Wang 1996, 175.

101   His own philosophical manifesto: Feferman 1990, II/176–187.

101   "I don't see any reason": "What Is Cantor's Continuum Problem?" (1964 supplement), in Feferman 1990, II/254–270, 268.

102   a previous essay: Paul Arthur Schilpp, ed., 1944, *The Philosophy of Bertrand Russell* (La Salle, Ill.: Open Court).

102   "in talking the matter over": Charles Parson, 1944, "Introductory Note to [Gödel] 1944," in Feferman 1990, II/102.

102   Russell's comment: Ibid., 102.

103   "It is not an altogether pleasant experience": Bertrand Russell, 1959, *My Philosophical Development* (London: Unwin Paperbacks), 159.

104 "My Philosophical Viewpoint": Wang 1996, 316.

105 "There are other worlds": Ibid., 288.

106 Gödel made his objections explicit: Kurt Gödel: "Some Observations About the Relationship Between Theory of Relativity and Kantian Philosophy," 1946/1949–C1, in Feferman 1995, III 230–259, 257–258 (note 27).

106 "a general feature of Kant's assertions": Kurt Gödel, 1961, "The Modern Development of the Foundations of Mathematics in the Light of Philosophy," in Feferman 1995, III/385.

107 "a procedure that should produce": Ibid., 383.

107 "[Husserl's] transcendental phenomenology": van Atten and Kennedy 2003, 470–471.

107 "German Idealism": Ibid., 443.

108 "what he is interested in": Roger Highfield and Paul Carter, 1993, The Private Lives of Albert Einstein (New York: St. Martin's Press), 225.

110 Almost all accounts: Paul Benacerraf, "Mathematical Truth," in Paul Benacerraf and Hilary Putnam, eds., 1983, Philosophy of Mathematics: Selected Readings (Cambridge: Cambridge University Press), 403–420, 403.

111 "I do not believe": Hilary Putnam, 1967, "Time and Physical Geometry," Journal of Philosophy 64, reprinted in Hilary Putnam 1979, Mathematics, Matter, and Method: Philosophical Papers, vol. I (Cambridge: Cambridge University Press), 198–205, 205.

111 time is "that mysterious": Kurt Gödel, A Remark About the Relationship Between Relativity Theory and Idealistic Philosophy," in Paul Arthur Schilpp, ed., 1949, Albert Einstein: Philosopher-Scientist (LaSalle, Ill.: Open Court), 557.

111 "continental" philosophers: Martin Heidegger, 1962, Being and Time, transl. J. Macquarrie and E. Robinson, (New York: Harper & Row); Edmund Husserl, 1990, On the Phenomenology of the Consciousness of Internal Time, trans. J.B. Brough (Dordrecht, Netherlands: Kluwer).

112 "now he has preceded me a little": Fölsing 1998, 741.

112 "the now means something special": Rudolf Carnap, "Intellectual Autobiography," in P.A. Schilpp, ed., The Philosophy of Rudolf Carnap (La Salle, Ill.: Open Court), 1963: 37–63, 37–38.

113 "Is what remains of temporal connection": Albert Einstein, "Reply to Criticisms," in Schilpp 1949, 687.

116 "Kurt Gödel's essay": Ibid., 687.

117–18 "Like most others": Lawrence Sklar, quoted in Palle Yourgrau, 1999, Gödel Meets Einstein: Time Travel in the Gödel Universe (Chicago: Open Court), xiii, note 14.

Chapter 7

120 Only when Gödel himself intervened: Dawson 1997, 184–185.

122 "The four-dimensional continuum is now": Albert Einstein,1961, Relativity: The Special and General Theory (New York: Corn Publishers), 149.

124 "time regained a real objective existence": James Jeans, "Man and the Universe," Sir Halley Stewart Lecture, 1936, in James Jeans et al., eds., 1936, Scientific Progress, (London: Allen & Unwin), 11–38, 22–23.

125 a famous philosophical essay on the A-series: A.N. Prior, 1959, "Thank Goodness That's Over," Philosophy 34, 12–17.

131 "Whether or not an objective lapse": Kurt Gödel, "A Remark about the Relation-ship Between Relativity Theory and Idealistic Philosophy," in Schilpp 1949, 562.

132 "The concept of existence": Ibid., 558, note 5.

132 "The notion of existence": Wang 1996, 150.

00   : Kurt Gödel, "A Remark about the Relationship Between Relativity Theory and Idealistic Philosophy," in Schilpp 1949, 558.

133 "As a substitute for absolute space": Kurt Gödel, 1949, "Lecture on Rotating Uni-verses," in Feferman et al. 1995, III/271.

136 "chronology protection conjecture": Hawking 1992.

137 "As we present time to ourselves": Wang 1996, 320.

137 "This concept of set": Gödel 1964, 258–259.

137 "These contradictions": Gödel 1961, 377.

140 one of the great texts in general relativity: Charles Misner, Kip Thorne, John Wheeler, Gravitation (New York: Worth), 1973.

140 "In physics . . . the possibility of knowledge": Gödel 1961, 377.

141 "the two great conceptual revolutions": Dyson 1993, 263.

141 "Gödel, he noted, "had taken down": J.A. Wheeler, Geons, Black Holes, and Quan-tum Foam: A Life in Physics (New York: W.W. Norton & Co.), 1998, 309–310.

141 "In a universe with an overall rotation": Wheeler 1998, 310.

142 an "unsettling consequence": Ibid., 344.

142 "the other thing that special relativity did": Ibid., 345.

142 Moreover, "every black hole": Ibid., 350.

143 "My, I wish we had talked to him": Ibid., 310.

Chapter 8

145 We live in a world in which ninety-nine per cent: Dawson 1997, 208.

145 Plato says: Wang 1996, 300.

145 "Geometry and Experience": Albert Einstein, 1921 "Geometry and Experience," in Albert Einstein, 1954, Ideas and Opinions (New York: Crown), 232–246.

146 "Einstein's now abandoned dream": John Wheeler, 1980, "Beyond the Black Hole," in Harry Woolf, ed., Some Strangeness in the Proportion (Reading, Mass.: Addison-Wesley), 344.

147 "One crazy man on the faculty": Dawson 1997, 194.

147 "You don't need it": Ibid., 195.

147 "I never go to lectures": Ibid., 203, note 5.

149 "I never thought he loved me so much": Levenson 2003, 430.

149 "My relationship to the Jewish people": Ibid., 428.

149 Simone Weil's dark study: Simone Weil, 1987, transl. A.F. Wills, The Need for Roots (London: Ark Paperbacks), 1987.

149 "the exaggerated esteem": Levenson 2003, 431.

149 "It is tasteless": Ibid., 432.

150 "Gödel's gnawing uncertainty": Ulam 1976, 80–81.

150 "Ulam doesn't understand my result": Wang 1996, 84.

151 "in 1905, work on spectral lines": Fölsing 1998, 224.

153 "I cannot define the real problem": Regis 1987, 33.

153 Gödel's contribution to the Schilpp volume: Gödel 1953/9, III & V, in Feferman Vol. III, 1995: 334–362.

153 "Some Basic Theorems on the Foundations of Mathematics, and Their Implications": Gödel, 1951, in Feferman et al. 1995, III/304–323.

154 "In view of widely held prejudices": Warren Goldfarb, "Introductory Note" to 1953/9, in Feferman et al. 1995, III/324.

155 "Ninety per cent of contemporary philosophers": Kurt Gödel, letter to Marianne Gödel, in Feferman et al. 2003, IV/437.

156 Willard V. Quine: Charles Parsons, 2002, "W.V. Quine: A Student's Eye View," *Harvard Review of Philosophy*, Spring, 2002, X:6–10, 8.

156–57 "One may wonder": Solomon Feferman, 1984, "Kurt Gödel: Conviction and Caution," *Philosophia Naturalis* 21, 1984: 546–562, 560.

157 "There are structural laws": Wang 1996, 151–152.

157 Christianity "best at the beginning": Ibid., 150.

157 "Since [Christ's] day there": John Hellman, 1982, *Simone Weil: An Introduction to Her Thought* (Wilfrid Laurier University Press), 60, 65.

158 "Philosophy tends to go down": Wang 1996, 150.

158 "Babylonian mathematics": Dawson 1997, 241, note 6.

158 "living corpse": Ibid., 234.

159 association with the logician Hao Wang: Wang 1996.

160 Einstein's ashes: Levenson 2003, 431.

Chapter 9

161 Engaging in philosophy is salutary: Wang 1996, 119.

162 "To date, only a single volume": Dawson 1997, 269.

165 "To do philosophy is a special vocation": Wang 1996, 308.

165 "Philosophy," he said: Ibid., 166.

165 "Actually, it would be easy": Ibid., 307.

165 The light dove: Kant 1965, A5, B9.

166 In a classic essay: Carl Hempel, "The Empiricist Criterion of Meaning," in A.J. Ayer, ed., 1959, *Logical Positivism* (New York: The Free Press), 108–129.

166 the first (and only) book: Wittgenstein, 1961.

167 "Philosophy must be of some use": F.P. Ramsey, 1931, ed. R.B. Braithwaite, *The Foundations of Mathematics and Other Logical Essays* (New York: Harcourt, Brace), 1931, 263.

167 "What we can't say we can't say": Ibid, 238.

167 material was provided for a lively book: David Edmonds and John Eidinow 2001, *Wittgenstein's Poker,* HarperCollins.

168 "After the devastating attacks": Paul Horwich, 1990, "The Growth of Now," Review of J.R. Lucas, *The Future, Times Literary Supplement*, June 22–28, 672.

168 entitled "Gödel's Philosophy": Warren Goldfarb, 1990, "On Gödel's Philosophy," address to the Association of Symbolic Logic, Helsinki, July 20, typescript.

169 entitled "Some Observations": Gödel 1946/49, B2, C1.

170 "the problem of time is important and difficult": Wang 1996, 319.

170 "The way . . . we form mathematical objects: Ibid., 301.

170 "a clarification of meaning": Gödel 1961, 383.

170 "focus more sharply on the concepts": Ibid.

170 "a new state of consciousness": Ibid.

171 "I don't particularly like Husserl's way": Wang 1996, 168.

171   "I love everything brief": Ibid., 43.

172   "entirely as intended by Kant": Ibid., 387.

172   in his path-breaking study *The Foundations of Arithmetic*: Frege 1980.

173   "Trying to see (i.e., understand) a concept": Wang 1996, 233.

173   "Often it is only after immense intellectual effort": Frege 1980, Introduction, vii.

173   "we perceive objects and understand concepts": Wang 1996, 235.

174   "We begin with vague perceptions": Ibid., 235.

174   "In arithmetic," wrote Frege: Frege 1980, 115.

174   "When the soul investigates by itself": Plato, 1981, transl. G.M.A. Grube, *Five Dialogues* (Indianapolis:, Hackett Publishing Company), *Phaedo*, 79d.

175   Both held that the only means of escaping: Palle Yourgrau, 1989, review essay, *Reflections on Kurt Gödel, Philosophy and Phenomenological Research* L, no. 2, 391–408, 402–403.

175   "Concepts are there," wrote Gödel: Wang 1996, 149.

175   "in the external world, in the whole of space": Frege 1980, 99.

176   *Constructibility and Mathematical Existence*: Charles Chihara, 1990, *Constructibility and Mathematical Existence* (Oxford: Clarendon Press).

176   "Chihara's account of Gödelian Platonism": E.P. James, 1992, "The Problem of Mathematical Existence," *Philosophical Books* XXXIII, no. 3, July, 129–138.

176   "mathematicians are prejudiced against intuition": Wang 1996, 169.

176   *Asymmetries in Time*: Paul Horwich, 1987, *Asymmetries in Time: Problems in the Philosophy of Science* (Cambridge, Mass.: The MIT Press).

177   Milič Čapek, another distinguished philosopher of science: Milič Čapek, 1966, "The Inclusion of Becoming in the Physical World," in Milič Čapek, ed., 1976, *The Concepts of Space and Time* (Dordrecht: Reidel).

177   Karl Popper had argued similarly: Karl Popper, 1982, *The Open Universe: An Argument for Indeterminism* (Totowa, New Jersey: Rowman and Littlefied), 2–3, note 2.

177   *The Disappearance of Time*: Palle Yourgrau, 1991, *The Disappearance of Time: Kurt Gödel and the Idealistic Tradition in Philosophy* (New York Cambridge University Press).

177   When an expanded edition appeared: Palle Yourgrau, 1999, *Gödel Meets Einstein: Time Travel in the Gödel Universe* (Chicago: Open Court).

177   *Bangs, Crunches, Whimpers, and Shrieks*: John Earman, 1995, *Bangs, Crunches, Whimpers, and Shrieks* (New York: Oxford University Press).

177   "the relative neglect of the philosophical moral": Ibid., 194–195.

177   "a deeply held conviction": Ibid, 195.

178   "this seems to me to be the correct response to Gödel": Ibid., 200.

179   "this is a pretty piece of ordinary language philosophizing": Ibid., 195.

180   "One may take the standpoint": in Schilpp 1949, 558, note 4.

182   "the publication of such an anthology": Wang 1987, 254.

182   "Even science," he said, "is very heavily prejudiced": Wang 1987, 254.

183   "One bad effect of logical positivism": Wang 1987, 308.

184   "It is given to us in its entirety": Wang 1996, 151.

184   "Mathematics is applied to the real world": Ibid., 151.

184   "Otherwise," he said, "mathematics would be just an ornament": Ibid.

WORKS CITED

Books

Beaney, Micheal, ed. 1997. *The Frege Reader*. Oxford: Blackwell Publishers.

Bell, E.T. 1986. *Men of Mathematics*. Touchstone.

Benacerraf, Paul and Hilary Putnam, eds. 1983. *Philosophy of Mathematics: Selected Readings*. Cambridge: Cambridge University Press.

Borges, Jorge Luis. 1964. *Labyrinths: Selected Stories and Other Writings*. A New Directions Book.

Bucky, Peter. 1992. *The Private Albert Einstein*. Kansas City: Andrews and McMeel.

Chihara, Charles S. 1990. *Constructibility and Mathematical Existence*. Clarendon Press.

Cornish, Kimberley. 1998. *The Jew of Linz*. London: Century.

Davis, Martin, ed. 1965. *The Undecidable*. New York: Raven Press.

Dawson, John. 1997. *Logical Dilemmas: The Life and Work of Kurt Gödel*. Wellesley, MA: A.K. Peters.

Dyson, Freeman. 1993. *From Eros to Gaia*. Penguin Books.

Earman, John. 1995. *Bangs, Crunches, Whimpers and Shrieks*. New York: Oxford University Press.

Edmonds, David and John Eidinow. 2001. *Wittgenstein's Poker*. New York: HarperCollins Publishers.

Einstein, Albert. 1954. *Ideas and Opinions*. New York: Crown Publishers, Inc.

———. 1961. *Relativity: The Special and General Theory*. New York: Corn Publishers.

Feferman, Solomon et al., eds. 1986. *Kurt Gödel: Collected Works, Vol. I*. New York: Oxford University Press.

———. 1990. *Kurt Gödel: Collected Works, Vol. II*. New York: Oxford University Press.

———. 1995. *Kurt Gödel: Collected Works, Vol. III*. New York: Oxford University Press.

———. 2003a. *Kurt Gödel: Collected Works, Vol. IV*. New York: Oxford University Press.

———. 2003b. *Kurt Gödel: Collected Works, Vol. V*. New York: Oxford University Press.

Fölsing, Albrecht (trans. Ewald Osers). 1998. *Albert Einstein*, Penguin Books.

Frank, Philipp. 1989. *Einstein: His Life and Times*. New York: Da Capo Press.

Frege, Gottlob. 1967. *Begriffsschrift, a formula language modeled on that of arithmetic, for pure thought*, in Jean van Heijenoort, ed. 1967: 1–82.

———. (trans. J.L. Austin). 1980. *The Foundations of Arithmetic: A Logico-Arithmetic Enquiry into the Concept of Number*. Evanston, IL: Northwestern University Press.

Gleick, James. 1992. *Genius: The Life and Science of Richard Feynman*. New York City, NY: Pantheon Books.

Hahn, Hans. 1980. *Empiricism, Logic, and Mathematics: Philosophical Papers*, edited by Brian McGuinnnes and D. Dordrecht. Reidel Publishing Company.

Halmos, Paul R. 1988. *I Want to Be a Mathematician: An Automathography in Three Parts*. Mathematical Association of America.

Heidegger, Martin (trans. J. Macquarrie and E. Robinson). 1962. *Being and Time*. New York: Harper & Row Publishers.

van Heijenoort, Jean, ed. 1967. *From Frege to Gödel: A Source Book in Mathematical Logic, 1879–1931*. Cambridge, MA: Harvard University Press.

Heisenberg, Werner. 1958. *Physics and Philosophy: The Revolution in Modern Science*. New York: Harper & Row Publishers.

———. 1983. *Encounters with Einstein, and Other Essays on People, Places, and Particles*. Princeton, NJ: Princeton University Press.

———. (trans. Peter Heath). 1990. *Across the Frontiers*. Woodbrige, Conn.: Ox Bow Press.

Hellman, John. 1982. *Simone Weil: An Introduction to Her Thought*. Wilfrid Laurier University Press.

Holton, Gerald and Yehuda Elkana, eds. 1982. *Albert Einstein: Historical and Cultural Perspectives*. Mineola, NY: Dover Publications, Inc.

Highfield, Roger and Paul Carter. 1993. *The Private Lives of Albert Einstein*. New York: St. Martin's Press.

Horwich, Paul. 1987. *Asymmetries in Time: Problems in the Philosophy of Science*. Cambridge, MA: The MIT Press.

Husserl, Edmund (trans. J.B. Brough). 1990. *On the Phenomenology of the Consciousness of Internal Time*. Dordrecht: Kluwer Academic Publishers.

Kant, Immanuel (trans. Norman Kemp Smith). 1965. *Critique of Pure Reason*. New York: St. Martin's Press.

Lenin, V.I. (trans. A. Fineberg). 2001. *Materialism and Empirio-Criticism*. London: 1930.

Levenson, Thomas. 2003. *Einstein in Berlin*. New York: Bantam Books.

Lindley, David. 2001. *Boltzman's Atom*. New York: The Free Press.

Menger, Karl. 1994. *Reminiscences of the Vienna Circle and the Mathematical Colloquium*. Dordrecht: Kluwer Academic Publishers.

Miller, Arthur. 1986. *Imagery in Scientific Thought*. Cambridge, MA: The MIT Press.

Misner, Charles W., Kip Thorne, and John Wheeler. 1973. *Gravitation*. Worth Publishers, Inc.

Monk, Ray. 1990. *Ludwig Wittgenstein: The Duty of Genius*. New York: The Free Press.

Moore, Walter. 1989. *Schrödinger: Life and Thought*. Cambridge: Cambridge University Press.

Nin, Anaïs. 1974. *A Spy in the House of Love*. Athens: Swallow Press.

Plato (trans. G.M.A. Grube). 1981. *Five Dialogues*. Indianapolis: Hackett Publishing Company.

Popper, Karl. 1982. *The Open Universe: An Argument for Indeterminism*. Totowa, NJ: Rowman and Littlefied.

Prieberg, Fred (trans. Christopher Dolan). 1991. *Trial of Strength: Wilhelm Furtwängler and the Third Reich*. Quartet Books.

Ramsey, F.P. 1931. *The Foundations of Mathematics and Other Logical Essays*, edited by R.B. Braithwaite. New York: Harcourt Brace.

Regis, Ed. 1988. *Who Got Einstein's Office?* Addison-Wesley Publishing Company, Inc.

Reid, Constance. 1986. *Hilbert–Courant*. New York: Springer-Verlag.

Rota, Gian-Carlo, 1997, *Indiscrete Thoughts*. Boston: Birkhaüser.

Russell, Bertrand. 1910–1913. *Principia Mathematica*. Cambridge: University Press.

———. 1959. *My Philosophical Development*. London: Unwin Paperbacks.

———. 2000. *Autobiography*. London: Routledge.

Saint Augustine (trans. R.S. Pine-Coffin). 1961. *Confessions*. Penguin Books.

Schilpp, Paul Arthur, ed. 1944. *The Philosophy of Bertrand Russell*. La Salle, IL: Open Court.

———, ed. 1949. *Albert Einstein: Philosopher-Scientist*. La Salle, IL: Open Court.

Schlick, Moritz. 1918. *Allgemeine Erkenntnislerhe*. Berlin: Springer.

de Spinoza, Benedict (trans. R. Elwes). 2001. *A Theologico-Political Treatise*. Blue Unicorn Editions.

Ulam, S.M. 1976. *Adventures of a Mathematician*. Berkeley: University of California Press.

Vonnegut, Kurt. 1999. *Mother Night*. Delta.

Wang, Hao. 1987. *Reflections on Kurt Gödel*. Cambridge, MA: The MIT Press.

———. 1996. *A Logical Journey: From Gödel to Philosophy*. Cambridge, MA: The MIT Press.

Weil, Simone (trans. A.F. Wills). 1987a. *The Need for Roots*. London: Ark Paperbacks.

———. 1987b. *Intimations of Christianity Among the Ancient Greeks*. London: Ark Paperbacks.

Weingartner, Paul and Leopold Schmetterer, eds. 1987. *Gödel Remembered*. Naples: Bibliopolis.

Wheeler, J.A. 1998. *Geons, Black Holes, and Quantum Foam: A Life in Physics*. New York: W. W. Norton & Co.

Whitrow, G.J. 1980. *The Natural Philosophy of Time*. Oxford: Clarendon.

Wittgenstein, Ludwig (trans. D.F. Pears and B. F. McGuinness). 1961. *Tractatus Logico Philosophicus*. London: Routledge & Kegan Paul.

———. (trans. Peter Winch). *Culture and Value*. Chicago: University of Chicago.

Yandell, Benjamin. 2002. *The Honors Class*. Natick, MA: A.K. Peters.

Yourgrau, Palle. 1991. *The Disappearance of Time: Kurt Gödel and the Idealistic Tradition in Philosophy*. New York: Cambridge University Press.

———. 1999. *Gödel Meets Einstein: Time Travel in the Gödel Universe*. Chicago: Open Court.

Articles

van Atten, Mark and Juliette Kennedy. 2004. "Gödel's Philosophical Development." *The Bulletin of Symbolic Logic* 9, no. 4 (December): 425–76.

Benacerraf, Paul. 1983. "Mathematical Truth." In Paul Benacerraf and Hilary Putnam, eds., *Philosophy of Mathematics: Selected Readings* (Cambridge: Cambridge University Press): 403–20.

Čapek, Milič. 1966. "The Inclusion of Becoming in the Physical World." In Milic Čapek, ed., *The Concepts of Space and Time* (Dordrecht: Reidel): 501–24.

Carnap, Rudolf. 1963. "Intellectual Autobiography." In P.A. Schilpp, ed., *The Philosophy of Rudolph Carnap* (La Salle, IL: Open Court): 37–63.

Dawson, John. 1984. "Discussion on the Foundations of Mathematics." In *History and Philosophy of Logic* 5: 111–29.

———. Edited and translated. 1988. "What Hath Gödel Wrought?" *Synthese* 114: 3–12.

Einstein, Albert. 1905a. "On the Electrodynamics of Moving Bodies." In *Annalen der Physik* 17: 891–921.

———. 1905b. "On the Movement of Particles Suspended in Fluids at Rest, as Postulated by the Molecular Theory of Heat." In *Annalen der Physik* 17: 549–60.

———. 1921. "Geometry and Experience." In Einstein, 1954: 232–46.

———. 1949. "Reply to Criticisms." In Schilpp, ed. (1949): 665–8.

Einstein, Albert, B. Podolsky and N. Rosen. 1935. "Can Quantum-Mechanical Description of Physical Reality be Considered Complete?" In *Physical Review* 47 (May 15): 777–80.

Feferman, Solomon. 1984. "Kurt Gödel: Conviction and Caution." In *Philosophia Naturalis* 21: 546–62.

Gödel, Kurt. 1944. "Russell's Mathematical Logic." In Feferman et al., eds. (1990): 119–41.

———. 1946. "Remarks before the Princeton Bicentennial Conference on Problems in Mathematics." In Feferman et al., eds. (1990): 150–53.

———. 1946/49—B2, C1. "Some Observations about the Relationship between Theory of Relativity and Kantian Philosophy." In Feferman et al., eds. (1995): 230–46; 247–59.

———. 1947. "What is Cantor's Continuum Problem?" (Supplement, 1964), in Feferman, et al., eds. (1990): 254–70

———. 1949a. "A Remark about the Relationship Between Relativity Theory and Idealistic Philosophy," in P. A. Schilpp, ed. (1949): 557–62.

———. 1949b. "Lecture on Rotating Universes." In Feferman et al., eds. (1995): 269–87.

———. 1951. "Some Basic Theorems in the Foundations of Mathematics and their Philosophical Consequences." The Gibbs Lecture, in Feferman et al., eds. (1995): 304–23.

———. 1953/59. "Is Mathematics Syntax of Language?" In Feferman et al., eds. (1995): 334–62.

———. 1961a. "The Modern Development of the Foundations of Mathematics in the Light of Philosophy." In Feferman et al., eds. (1995): 375–87.

———. 1961b. Letter to Marianne Gödel. In Feferman et al., eds. (2003a): 435–37.

———. 1968. Letter to Hao Wang. In Feferman et al., eds. (2003b): 403–05.

———. 1970. "Ontological Proof." In Feferman et al., eds. (1995): 403–04.

———. 1973. Letter to Abraham Robinson, in Feferman et al., eds. (2003b): 201.

Goldfarb, Warren. 1996. "The Philosophy of Mathematics in Early Positivism." In R.N. Giere, ed., *Origins of Logical Empiricism* (Minnesota Studies in the Philosophy of Science: University of Minnesota Press) 16: 213–30.

———. 1990. "On Gödel's Philosophy." Association of Symbolic Logic Address (Helsinki, July 20), unpublished typescript of address.

———. 1995a. "Introductory Note to 1953/9." In Feferman et al., eds. (1995).

———. 1995b. "On Gödel's General Philosophical Outlook." Boston University Conference: "Gödel's General Philosophical Significance" (February 6–7) unpublished address.

Hempel, Carl. 1959. "The Empiricist Criterion of Meaning." In A.J. Ayer, ed., *Logical Positivism* (New York: The Free Press, 1959): 108–29.

Holton, Gerald. 1998. "Einstein and the Cultural Roots of Modern Science." *Daedalus* (Winter): 1–44.

Horwich, Paul. 1990. "The Growth of Now." Review of J.R. Lucus, *The Future. Times Literary Supplement* (June 22–28): 672.

James, E.P. 1992 "The Problem of Mathematical Existence." *Philosophical Books* XXXIII, no. 3 (July): 129–38.

Jeans, James. 1936. "Man and the Universe." Sir Halley Stewart Lecture. In James Jeans et al., eds., *Scientific Progress* (London: Allen and Unwin, 1936): 11–38.

Kreisel, Georg. 1980. "Kurt Gödel: 1906–1978." *Biographical Memoirs of Fellows of the Royal Society* 26: 149–224.

Malament, David. 1984. "Time Travel in the Gödel Universe." *Proceedings of the Philosophy of Science Association* 2: 91–100.

Overbye, Dennis. 1999. "Einstein, Confused in Love and, Sometimes, Physics." *The New York Times* (August 31).

Parsons, Charles. 1990. "Introductory Note to [Gödel] 1944." In Feferman (1990): 102–18.

———. 2002. "W.V. Quine: A Student's Eye View." *Harvard Review of Philosophy* (Spring) X: 6–10.

Post, Emil. 1965. "Absolutely Unsolvable Problems and Relatively Undecidable Propositions: Account of an Anticipation." In Martin Davis, ed., *The Undecidable: Basic Papers on Undecidable Propositions, Unsolvable Problems and Computable Functions* (New York: Raven Press, 1965): 340–433.

Prior, A.N. 1959. "Thank Goodness That's Over." *Philosophy* 34: 12–17.

Putnam, Hilary. "Time and Physical Geometry." In *Mathematics, Matter, and Method: Philosophical Papers 1* (Cambridge: Cambridge University Press, 1979): 198–205.

Russell, Bertrand. 1902. "Letter to Frege." In van Heijenoort, ed. (1967): 124–25.

Sklar, Lawrence. "Comments on Malament's 'Time Travel in the Gödel Universe.'" *Proceedings of the Philosophy of Science Association* 2 (1984): 106–10.

———. 1999. In Palle Yourgrau *Gödel Meets Einstein: Time Travel in the Gödel Universe* (Chicago: Open Court, 1999) XIII, no. 14.

Suber, Peter. 1992. "50 Years Later, The Questions Remain: Kurt Gödel at Blue Hill." *Ellsworth American* (August 27).

Wang, Hao. 1995. "Time in Philosophy and in Physics." *Synthese* 102: 215–34.

Wheeler, John. 1980. "Beyond the Black Hole." In Harry Woolf, ed., *Some Strangeness in the Proportion* (Addison-Wesley Publishing Company, 1980).

Wigner, Eugene, 1979. "The Unreasonable Effectiveness of Mathematics in the Natural Sciences." In *Symmetries and Reflections: Scientific Essays* (Woodbridge, CT: Ox Bow Press): 222–37.

Yourgrau, Palle. 1989. "Review Essay: *Reflections on Kurt Gödel*," *Philosophy and Phenomenological Research* L, no. 2: 391–408.

INDEX